View of general offices and laboratories, American Institute of Laundering, National Trade Association of the Laundry Industry, Joliet, Illinois.

FUNDAMENTALS
of
DETERGENCY

by

William W. Niven, Jr.
Associate Research Chemist
Midwest Research Institute

Under the sponsorship of the
American Institute of Laundering,
National Trade Association of the Laundry Industry

Reinhold Publishing Corporation
1950

Second Printing, 1951

PRINTED IN THE UNITED STATES OF AMERICA
BY AMERICAN BOOK–KNICKERBOCKER PRESS, INC., NEW YORK

Preface

Most technical references on detergency are highly specialized and assume that the reader is fully conversant with the basic theoretical chemistry of the subject. As specialized treatises they are excellent. However, they do not, individually, fill an acute need for a comprehensive reference from which anyone with a reasonable amount of technical training can obtain a complete picture of the various ramifications of detergency. For example, a technical article on detergency may discuss the detergent action of soap as related to surface activity, with only the briefest mention as to *how* soap affects surface or interfacial tension. When one turns then to a physical chemistry text to determine the *how*, he may find a brief explanation as to how soaps in general are surface-active but no mention of the effects of such factors as composition, temperature, concentration, and pH. It then becomes necessary to turn to still other specialized papers. The author is qualified to testify that such a course of pursuit of information can be both devious and time-consuming.

From the above, the original purpose of the present dissertation is obvious, namely, to bridge the gap between general and specialized information on detergency. A further purpose arose spontaneously as work on the book progressed. Thus, in the somewhat mechanical process of compiling, assembling, and correlating reference material, certain gaps in the basic knowledge have been revealed and certain inconsistencies in published data have become apparent. It is hoped that the present work will serve usefully to some extent to suggest and to guide future studies.

An attempt has been made first to set forth a discussion of the fundamentals of detergency and then to apply these fundamentals to a practical detergent process. Laundering has been selected as the illustration of a practical application, not only in view of the sponsorship of this work but also because there probably is no other single application which involves so completely all phases of detergency.

A detailed listing and discussion of the hundreds of synthetic surface-active agents now available on the market has been avoided purposely. In the first place, it would have occupied a large portion of the book and there are excellent reference sources for such information. In the second place, the discussions in the book are intended to apply, in so far as possible, to detergency and detergent actions and not to specific detergent compounds. Where mention is made of a specific detergent, it is given to illustrate the effect of composition on some property or action.

Certainly in a book which is as much in the nature of a review as the present one, first acknowledgment is due those from whom data and *reasoning* have been borrowed so extensively. It has been possible to give specific mention only to a relatively few of the many hundreds of authors of the literature of detergency. Whether listed or not, all the references have been most helpful.

Special acknowledgment is due the American Institute of Laundering for sponsorship of the work and for many helpful discussions, especially those relative to applications to commercial laundering.

The manuscript for Part I of the book has been carefully read and criticized by Dr. Loren Morey. His many suggestions as to improvement in form of presentation also are gratefully acknowledged. Discussions of the chemistry of soaps and synthetic detergents have been reviewed by Mr. K. R. Hoffman. Editing of the complete manuscript was by Miss Sarah Lechtman. Substantially all drawings were prepared by Mr. J. K. Schweppe.

Finally, acknowledgment is due particularly to Miss Jane Hathaway, without whose assistance in the extracting of references from the literature, assembling of data, *reasoning out* of technical points, and accomplishment of the multitudinous tasks related to preparation of the final manuscript, this book might not have been completed.

WM. W. NIVEN, JR.

Kansas City, Missouri

Contents

PART I

FUNDAMENTAL CONSIDERATIONS OF THE DETERGENT PROCESS

"*What is fundamental research today is applied research tomorrow.*"

— Lammot du Pont

1

Introduction and Historical Review

The term **deterge** denotes "to cleanse" or "to purge away." **Detergence** and **detergency** denote "cleansing quality" or "cleansing power." In turn, **detergent** is used as an adjective to mean "cleansing" or "purging" or as a noun to denote "a cleansing agent." It is evident from these definitions that, in the broadest sense, detergency applies to any process wherein foreign or undesirable material is removed from a desirable material, whether the latter be a solid, such as clothing or dishes, or even a liquid or a gas. In the practical, more limited sense, however, we think of detergency as pertaining only to the removal of undesirable material from solids. The application is still broad, even with this limitation. Thus, we have detergency in laundering, in the manufacture of textiles, in the preparation of metals for processing, and in a multitude of housekeeping "chores."

The discussions presented in this volume are primarily related to detergency in laundering. It is not meant by this that the *basic* principles of detergency in laundering are any different from those of detergency in dish washing or in any other cleaning process. Most of what may be said concerning detergency in one case applies equally well in other cases. However, there are special considerations peculiar to each particular application. The reader interested in some application other than laundering should be readily able to detect those special considerations which do not fit his particular interest.

The identity of the first maker of soap is shrouded in anonymity but he certainly lived in very ancient times. Perhaps the accidental spilling of fat on hot wood ashes produced the first soap. Plinius reported in about 70 A.D. that soap was prepared at that time by boiling fat with wood ashes and lime. The earliest developments and improvements of soap were either accidental or the fruit of constant "trial and error" experimentation, for it was not until 1741 that the

Figure 1–1. Modern soap production. Soap flakes coming from the dryer. (Photograph courtesy The Procter & Gamble Company.)

first light was shed, by Geoffroy, on the constitution of this material. He observed that the fatty material obtained on decomposing soap with mineral acid differed in properties from the fat from which the soap was made. Unfortunately, Geoffroy's work became lost in obscurity, and it was not until 1811, when Chevreul,[1] the "father of oil and fat chemistry," made his famous analysis of soap, that the real foundation of soap chemistry was formed. Thus, Chevreul showed that soap consisted of the alkali salts (sodium or potassium, as the case might be) of fatty acids, as obtained by saponification of fats and oils.[2]

The well-known property of soaps to form insoluble salts with calcium or magnesium hardness in water and their susceptibility to acid decomposition have prompted the development of synthetic detergents, so designed molecularly as to overcome these deficiencies. The first of these "synthetics" was a soap of sulfated castor oil, developed about 1860 as a textile assistant. Since that time, and particularly in recent years, the development of synthetic detergents of several types has taken enormous strides so that today there are literally hundreds of such materials on the market.

The art of laundering is probably older than the art of soapmaking. There are still those today who launder their clothes down at the creek without benefit of any "detergent" other than a stick, a stone, and "elbow grease." One of the earliest written instructions on the art of laundering was "The Family Dyer and Scourer," by William Tucker, published in London in 1817. This treatise recognized the advantage of the use of alkaline builder with soap and also gave instructions for the use of bluing to produce a "brilliant white."[3] On the other hand, laundering as an industry is relatively a newcomer in the general field of cleansing. Laundering became a business in this country during the California gold rush when one disillusioned miner turned laundryman to supply the needs of the exclusively male population of the gold fields. By 1851, the Contra Costa Laundry was operating in Oakland, California.[4] From such an insignificant start, the laundry industry has grown to the stature of an $800,000,000 annual volume in this country alone.

It was not until the end of the 19th Century that concerted attention was given to the ways in which soap solutions accomplish

---

[1] Chevreul, "Recherches chimiques sur les corps gras d'origine animale," Paris, 1823.
[2] Lewkowitsch, J., *J. Soc. Chem. Ind.*, **26**, 590–3 (1907).
[3] Harwood, F. C., *J. Textile Inst.*, **39**, 513–25 (1948).
[4] Kramer, V., and Huegy, H. W., "Establishing and Operating a Laundry," Industrial (Small Business) Series No. 37, U.S. Department of Commerce, Government Printing Office, Washington (1946).

their excellent work of removing soil. The works of Krafft, Donnan, Hillyer, Jackson, Spring, and others of that period might be considered to signify the birth of detergency as a truly scientific subject. The enormous amount of study since that time attests to both the importance and the complexity of the subject. The ways in which our present knowledge of detergency has been developed from this start will be apparent from the discussions that follow.

2

Nature and Properties of Detergents

Soaps

The basic principles by which most soap is made have not changed materially for hundreds of years; however, refinements in materials, methods, and equipment have resulted in soaps so improved as to be almost new products. The greater amount of soap is produced by alkali saponification of animal and vegetable fats, the glycerides of long-chain fatty acids. Basically, the saponification reaction may be illustrated as

$$\underset{\text{(sodium hydroxide)}}{3NaOH} + \underset{\text{(glyceryl stearate)}}{(C_{17}H_{35}COO)_3C_3H_5} \rightarrow \underset{\text{(sodium stearate soap)}}{3C_{17}H_{35}COONa} + \underset{\text{(glycerine)}}{C_3H_5(OH)_3} \quad (1)$$

using glyceryl stearate (one of the principal constituents of tallow) as a typical example. The methods of obtaining the soap thus formed in a solid state separate from the glycerine are not within the scope of the present discussion.

Recently a somewhat different process has attained importance in the industry. The reactions of this process, involving first the hydrolysis of the fat to fatty acid and glycerine and then the neutralization of the fatty acid with soda ash, may be represented as follows:

$$\underset{\text{(glyceryl stearate)}}{(C_{17}H_{35}COO)_3C_3H_5} + 3H_2O \rightarrow \underset{\text{(stearic acid)}}{3C_{17}H_{35}COOH} + \underset{\text{(glycerine)}}{C_3H_5(OH)_3} \quad (2)$$

$$\underset{\text{(stearic acid)}}{2C_{17}H_{35}COOH} + \underset{\text{(soda ash)}}{Na_2CO_3} \rightarrow \underset{\text{(sodium stearate soap)}}{2C_{17}H_{35}COONa} + CO_2 + H_2O \quad (3)$$

Tallow constitutes approximately 40 per cent of all the fats and oils consumed by the soap industry.[1] *Coconut oil* is second, representing about 25 per cent of the total. Other significant sources are *palm kernel oil*, *palm oil*, *fish oils*, and *greases* from hogs and garbage. All of these fats and oils are mixtures of several glycerides, with one generally predominating, as shown in Table 2–1. All commercial soaps are

TABLE 2–1. PRINCIPAL FATTY ACIDS FROM FATS AND OILS USED FOR SOAP [2]

Approx. % Acid	Olive Oil	Palm Oil	Palm Kernel Oil	Coconut Oil	Ox Tallow	Sheep Tallow	Pig Fat
Caprylic			4	5–10			
Capric			5	7			
Lauric			40–50	45			
Myristic		2	15–30	18	5	5	1
Palmitic	7–15	35–45	7	10	30	25	30
Stearic	1–3	5	2	2	15–25	30	12
Oleic	70–85	40–50	10–15	7	40–50	36	48
Linoleic	5–15	10	1				6

mixtures of more than one fatty acid salt, not only because of the complexity of separating the original fats into pure glycerides but also because of the frequently more desirable detergent properties of mixed soaps, as will be evident from later discussions.

The properties of a commercial soap depend upon the properties of the individual fatty acid salts of which it is composed. Briefly, the fatty acids of most consequence in soapmaking are as indicated in Table 2–2. Actually, the salts of fatty acids below C_{10} or C_{12} have only weak "soap-like" properties.

TABLE 2–2. FATTY ACIDS COMMONLY USED FOR SOAPS

Acid	Formula	Melting Point, °C [3]
	Saturated Fatty Acids	
Caproic	$C_6H_{12}O_2$	−1.5
Caprylic	$C_8H_{16}O_2$	16
Capric	$C_{10}H_{20}O_2$	31.5
Lauric	$C_{12}H_{24}O_2$	48(44)
Myristic	$C_{14}H_{28}O_2$	57–8
Palmitic	$C_{16}H_{32}O_2$	63–4
Stearic	$C_{18}H_{36}O_2$	69–70
	Unsaturated Fatty Acids	
Oleic	$C_{18}H_{34}O_2$	14
Linoleic	$C_{18}H_{32}O_2$	−11

[1] Shreve, R. N., "The Chemical Process Industries," p. 601, New York, McGraw-Hill Book Co., Inc., 1945.

[2] Prepared from values given by Bailey, A. E., "Industrial Oil and Fat Products," New York, Interscience Publishers, Inc., 1945.

[3] Lange, N. A., "Handbook of Chemistry," 5th ed., Sandusky, Ohio, Handbook Publishers, Inc., 1944.

Figure 2–1. Working floor of a soap kettle house. (Photograph courtesy The Procter & Gamble Company.)

To illustrate the extent to which a commercial soap produced from a given fat or oil may be a mixture of several soaps, Table 2–1 shows typical compositions of the fatty-acid fraction obtainable from the more important vegetable and animal fat sources. Fatty acids present in only very small amounts are not included. No attempt is made to indicate in all cases the extent to which the percentages may vary between samples. Fish oils, also used to an appreciable extent in soapmaking but not included in the table, contain large proportions of highly unsaturated fatty acids which must be hydrogenated for use in soaps.

The relative properties of alkali salts of fatty acids are those of a typical homologous series of compounds. Thus, at a given temperature the solubilities of salts decrease with increase in length of the carbon chain. To attain a concentration of higher salts in water of, say, 0.2 per cent as might be used in laundering, the temperature must be increased with increasing length of the carbon chain. Likewise, the physical hardness and melting point of salts increase with greater length of carbon chain. The presence of unsaturation in the carbon chain, as in the case of sodium oleate, results in a "reduction" of the physical properties to compare with those of a saturated fatty-acid salt of fewer carbon atoms. The homology of the series of fatty-acid salts applies not only to the fatty portion of the soap molecules but also to the alkali portion. Thus, a series of soaps using the same fatty acid but varying in alkali constituent in the order lithium, sodium, potassium, rubidium, and cesium would exhibit typical homologous relationships. Of these alkalies, however, sodium is used by far to the greatest extent, potassium to some extent, and the others hardly at all except for special purposes. Potassium (potash) soaps are in general softer and more soluble than the corresponding sodium soaps.

The acid character of the fatty acid constituent of soap is not limited to reaction with alkalies. Soaps of alkaline earths, such as calcium and magnesium, and of heavy metals, such as lead and cobalt, are common but are not detergents because of their very low solubilities. In fact, the alkaline earth soaps are not only insoluble but are excellent water repellents. Some heavy metal soaps are used as driers in paints and varnishes.

Synthetic Detergents

The instability of soap in acid solutions and the insolubility of

Figure 2–2. Modern plant for converting fats to fatty acids. (Photograph courtesy The Procter & Gamble Company.)

alkaline earth soaps provided the stimulus for the development of synthetic detergents. The first of these was the sodium salt of sulfated castor oil (Turkey Red oil).

Fieser and Fieser[4] classify synthetic surface-active agents into five general types, according to their chemical composition. These types, some of which are good detergents and all of which are wetting agents, are set forth below.

TYPE 1. *Sulfates or sulfonates of long-chain alcohols.* These agents are in turn divided into three groups:

1. Those formed by the general reaction:

$$\underset{\text{(fatty acid)}}{RCOOH} \rightarrow \underset{\text{(alcohol)}}{RCH_2OH} \xrightarrow[\text{neutralization}]{\text{sulfation and}} \underset{\text{(sodium alkyl sulfate)}}{RCH_2OSO_2ONa} \quad (4)$$

These agents are resistant to hard water, stable at low pH, and are good wetting agents and detergents.

2. Secondary alcohol sulfonates, such as "Aerosol OT," formed from maleic anhydride, an appropriate long-chain alcohol, and sodium bisulfate. These compounds are primarily wetting agents.

3. Alkylaryl sodium sulfonates of the type $R \cdot Ar \cdot SO_3Na$. These compounds are derived from petroleum at a cost which compares favorably with that of soap.

TYPE 2. *Compounds made resistant to hard water by blocking of the carboxyl group.* Two compounds are representative of this type. The first, "Igepon A," is prepared as follows:

$$\underset{\text{(oleic acid)}}{CH_3(CH_2)_7CH{=}CH(CH_2)_7COOH} + \underset{\text{(isethionic acid)}}{HOCH_2CH_2SO_3H} \rightarrow$$

$$\underset{\text{("Igepon A")}}{CH_3(CH_2)_7CH{=}CH(CH_2)_7COOCH_2CH_2SO_3Na} \quad (5)$$

Due to hydrolysis at the ester grouping, this compound is not stable in alkaline solution. This condition is overcome by means of an amide linkage in the second compound of this type, which is formed as follows:

$$\underset{\text{(oleic acid)}}{CH_3(CH_2)_7CH{=}CH(CH_2)_7COOH} + \underset{\text{(taurine)}}{NH_2CH_2CH_2SO_3H} \rightarrow$$

$$\underset{\text{("Igepon T")}}{CH_3(CH_2)_7CH{=}CH(CH_2)_7CO{-}NHCH_2CH_2SO_3Na} \quad (6)$$

---

[4] Fieser, L. F., and Fieser, M., "Organic Chemistry," pp. 395–7, Boston, D. C. Heath and Company, 1944.

TYPE 3. *Compounds with nonionizing hydrophilic groups.* These compounds utilize nonionizing hydrophilic groups to attain stability in hard water and may be divided into three groups, as follows:

1. Esters of polyglycerol and a fatty acid. Glycerol, condensed to pentaglycerol and reacted with a fatty acid, yields pentaglycerol monoester. The unesterified hydroxyl groups serve as the water-attracting groups; several are required because the hydroxyl group is less hydrophilic than sulfate, carboxyl, or sulfonate groups found in other compounds.

2. A second group consists of glycol esters of fatty acids, derived from treatment of the acid with ethylene oxide.

3. The third group is formed in the following manner:

$$\underset{\text{(urea)}}{OC\begin{matrix} \diagup NH_2 \\ \diagdown NH_2 \end{matrix}} \xrightarrow[\text{(ethylene oxide)}]{\overset{CH_2-CH_2}{\diagdown O \diagup}} OC\begin{matrix} \diagup NHCH_2CH_2OH \\ \diagdown NHCH_2CH_2OH \end{matrix} \xrightarrow[\text{(fatty acid)}]{RCOOH} OC\begin{matrix} \diagup NHCH_2CH_2OCOR \\ \diagdown NHCH_2CH_2OCOR \end{matrix} \quad (7)$$

TYPE 4. *Triethanolamine soaps.* Triethanolamine salts of fatty acids are purely organic soaps. They are excellent emulsifiers and good dry-cleaning soaps. Their cost is high, compared to that of ordinary soaps.

TYPE 5. *Invert soaps or cation-active soaps.* These compounds are so named because the organic portion of the molecule forms a positively charged ion in solution, instead of a negatively charged ion as with ordinary soap. They cannot react with heavy metal ions and are used in neutral or acid solution. They have marked bactericidal properties but their use is limited because of high toxicity. An example of such a compound, a Sapamine from oleic acid, has the following general structure:

$$[C_{17}H_{33}CONHCH_2CH_2N^+(CH_3)_3]_2SO_4^{=}$$

It should be borne in mind that the above-mentioned agents are listed only as being typical of their particular type. There are literally hundreds of surface-active agents which already have been developed, and there are certainly many more yet to come. Of the types of agents listed, probably the one of greatest interest to laundering at the present time is Type 1, the sulfates and sulfonates of long-chain alcohols. Although these compounds have not yet been extensively accepted in commercial laundry work, developments in that direction are going forward rapidly. Consequently, most of the discussions of synthetic detergents in subsequent chapters will be directed to this type of compound. Many of the agents of other types have been designed for highly specialized applications in textile manufacturing but they have not been extensively studied in relation to their applicability to laundering. For more detailed discussions and classifications of synthetic surface-active agents, the reader is referred to recent papers by Dreger and coworkers,[5] by McCutcheon[6] and to Chapter 3, following.

---

[5] Dreger, E. E., Keim, G. I., Miles, G. D., Shedlovsky, L., and Ross, J., *Ind. Eng. Chem.*, **36**, 610–17 (1944).

[6] McCutcheon, J. W., *Chem. Industries*, **60**, 811–24 (1947).

3

Chemistry of Unbuilt Detergent Solutions

Hydrolysis in Detergent Solutions

Of first importance in a discussion of the physical and chemical nature of aqueous soap solutions is the susceptibility of soap to hydrolysis. Being salts of very weakly ionizing fatty acids, hydrolysis is probably the most significant chemical reaction into which soaps enter. On the other hand, synthetic detergents wherein hydrolysis has been structurally blocked exhibit hydrolysis only under extreme conditions. Thus, the anionic synthetic detergents are in reality salts of strong acids.

In very dilute aqueous solution, soaps and ionic detergents in general can be expected to act as simple fairly strong electrolytes. Electrolytic dissociation of these materials can be illustrated by the following examples:

$$\underset{\text{(sodium alkyl sulfate)}}{RCH_2OSO_2ONa} \rightleftharpoons Na^+ + \underset{\text{(alkyl sulfate anion)}}{RCH_2OSO_2O^-} \qquad (1)$$

$$\underset{\text{(sodium soap)}}{RCOONa} \rightleftharpoons Na^+ + \underset{\text{(fatty anion)}}{RCOO^-} \qquad (2)$$

In addition, soaps will hydrolyze in dilute solution in the following manner:

$$RCOO^- + H_2O \rightleftharpoons \underset{\text{(fatty acid)}}{RCOOH} + OH^- \qquad (3)$$

In all of the above cases the extent to which the reaction takes place will depend on such factors as length and degree of saturation of the paraffin chain, R, and on the temperature and concentration. These controlling factors will be considered in detail below.

Krafft and Wiglow[1] claimed in 1895 that soap could be completely hydrolyzed in the manner illustrated by combined equations (2) and (3) if the free fatty acid were removed by a solvent. However, Lewkowitsch[2] has shown experimentally that even with extraction of the fatty acid, free alkali remaining will eventually attain such a concentration that the practical limit is reached for extraction of the fatty acid and, thus, for completion of the hydrolysis.

One of the earliest studies of soap hydrolysis was that of Krafft and Stern[3] in 1894, the chief significance of which for us at the present time is the postulation of the formation of acid soaps as well as of free fatty acid by hydrolysis. They considered that the formation of acid soap takes place by combination of the free fatty acid with neutral soap remaining undecomposed in solution. Although the stoichiometry of this reaction is questionable, their postulation may be schematically illustrated as

$$\underset{\text{(fatty acid)}}{x\text{RCOOH}} + \underset{\text{(soap)}}{y\text{RCOONa}} \rightarrow \underset{\text{(acid soap)}}{(\text{RCOOH})_x(\text{RCOONa})_y} \qquad (4)$$

The actual products of hydrolysis of a soap at a given temperature are a function of the soap concentration. The process illustrated by equation (3) to yield free fatty acid can be expected to be the predominant reaction in very dilute solution, that is, at concentrations where the amount of fatty acid formed does not exceed its solubility in water. The dependency of this limiting concentration for "normal" hydrolysis on the length of the paraffin chain is well illustrated by the data in Table 3–1.

The complete process of hydrolysis in more concentrated solutions, wherein acid soaps are formed, is by no means as simple as indicated by equation (4). Perhaps the most comprehensive attempt to follow the changes in composition of a soap solution with changes in concentration is that of Ekwall and Lindblad,[4] who followed the changes electrometrically in sodium laurate solution at 20°C, through

---

[1] Krafft, F., and Wiglow, H., *Ber.*, **28**, 2566–73 (1895).
[2] Lewkowitsch, J., *J. Soc. Chem. Ind.*, **26**, 590–3 (1907).
[3] Krafft, F., and Stern, A., *Ber.*, **27**, 1747–61 (1894).
[4] Ekwall, P., and Lindblad, L. G., *Kolloid-Z.*, **94**, 42–57 (1941).

TABLE 3–1. FATTY ACID FROM SOAP HYDROLYSIS [6]

Soap Concentration, (Molar)	Hydrolysis,[a] Per cent	Fatty Acid Concentration, (Molar)
Sodium Laurate (60°)		
0.0001	0.8	0.8×10^{-6}
.001	.3	3.0×10^{-6}
.005[b]	.25	12.0×10^{-6}[c]
.01	.3	34×10^{-6}
.03	.4	150×10^{-6}
.1	.4	370×10^{-6}
Sodium Myristate (60°)		
0.0001	1.8	1.8×10^{-6}
.0003	1.4	3.6×10^{-6}
.001[b]	1.2	11.7×10^{-6}[c]
.005	1.75	125×10^{-6}
.01	1.75	170×10^{-6}
Sodium Palmitate (60°)		
0.00001	18	0.2×10^{-6}
.0001	4	3.8×10^{-6}
.0002[b]	3.8	7.6×10^{-6}[c]
.0003	3.8	13.2×10^{-6}
.001	11	130×10^{-6}
.003	17	540×10^{-6}
Sodium Stearate (60°)		
0.000002	40	0.8×10^{-6}
.00003[b]	8	4.0×10^{-6}[c]
.0001	22	25×10^{-6}
.001	46	470×10^{-6}
.01	13	1600×10^{-6}

[a] By interpolation from graphs of Powney and Jordan.
[b] Concentration at which hydrolysis commences to increase to a maximum.
[c] Approximate limit of solubility of fatty acid in water.

the range of $0.001N$ to $0.18N$. Their results, as shown in Figure 3–1, are interpreted by them according to Ekwall's theory of hydrolysis[5] as follows:

The limiting sodium laurate concentration for normal hydrolysis in the manner illustrated by equations (2) and (3) to form free lauric acid is $0.006N$. Below this concentration, where the solution is unsaturated with respect to lauric acid, the degree of hydrolysis decreases in a normal manner as the concentration increases, that is, the decrease is the result of the mass action effect on the equilibrium of the hydrolysis reaction.

[5] Ekwall, P., *Kolloid-Z.*, **92**, 141–57 (1940).

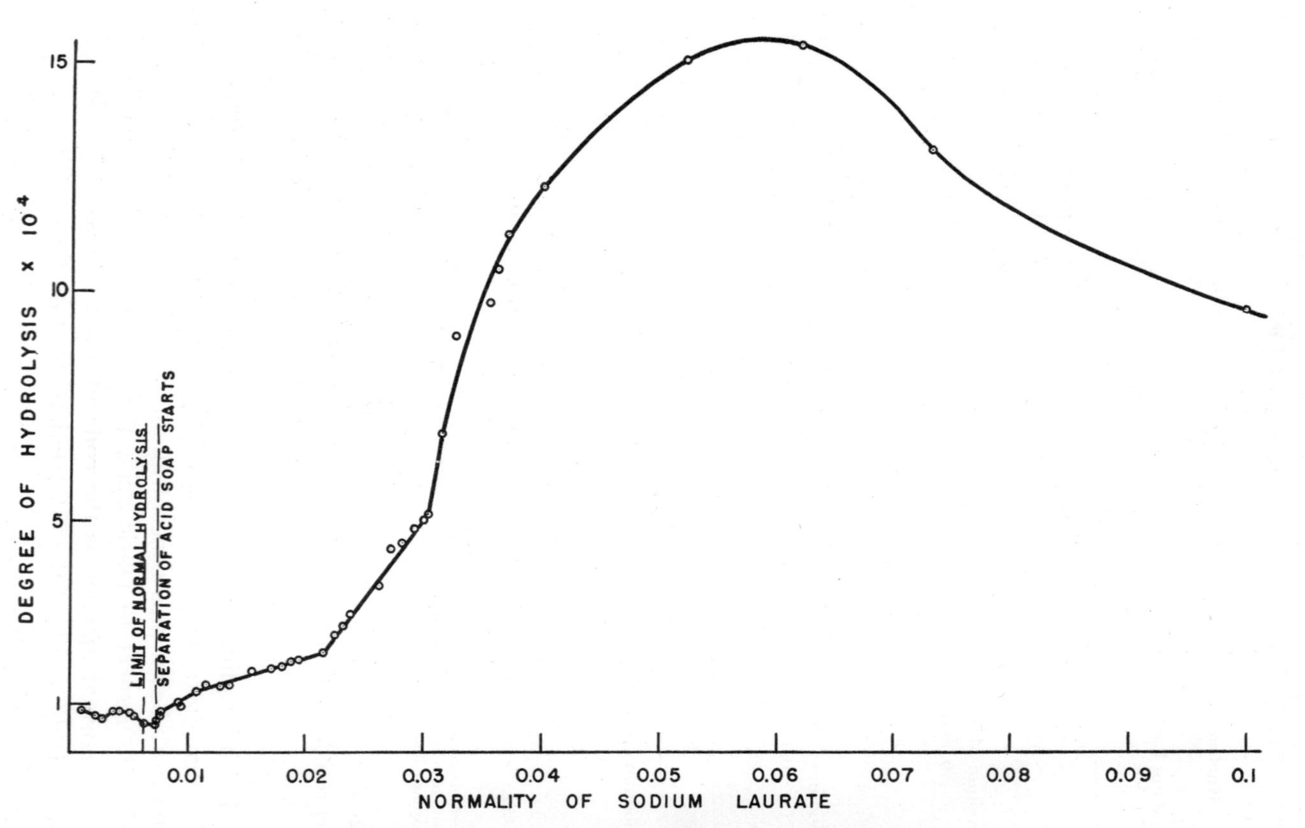

Figure 3–1. The hydrolysis of sodium laurate as a function of concentration, at 20°C.[4] ("Kolloid Zeitschrift" is published by Theodor Steinkopff, of Dresden and Leipzig, Germany. The German interests in the U.S. Copyright in this periodical were vested in 1946 by the Alien Property Custodian. The reproduction in this book of the data shown on Figure 3–1 is by permission of the Attorney General of the United States in the public interest under License No. JA–1398.)

Immediately beyond the limiting concentration of 0.006*N*, the degree of hydrolysis again falls a little and the separated free lauric acid diminishes in quantity. This is a narrow region, before *acid soap* begins to separate, where the solution is clear and in which hydrolysis yields chiefly the ion $(LHL)^-$, L denoting the laurate group, as follows:

$$\text{Na}\cdot\text{Laurate} + \text{HOH} \rightleftharpoons \text{Na}^+ + \text{OH}^- + \underset{\text{(lauric acid)}}{\text{H}\cdot\text{Laurate}}$$

$$\text{H}\cdot\text{Laurate} \rightleftharpoons \text{H}^+ + (\text{Laurate})^-$$

$$\text{H}\cdot\text{Laurate} + (\text{Laurate})^- \rightleftharpoons (\text{H}(\text{Laurate})_2)^- \tag{5}$$

There follows then an intermediate range of soap concentration, above 0.0075*N*, in which the degree of hydrolysis increases rather than decreases with rising concentration. This is the range where acid soap separates from the solution, the separation thus favoring an increase in the degree of hydrolysis.

Ekwall pictures the reaction of the doubled fatty ion of equation (5) to form acid soap in accordance with the following:

$$(\text{H}(\text{Laurate})_2)^- + \text{Na}^+ \rightarrow \text{NaH}(\text{Laurate})_2 \tag{6}$$

This is in contrast to Krafft's[3] theory of acid soap formation wherein the simple ions of fatty acid are considered to hydrolyze to fatty acids and then produce acid soaps by secondary reaction with undecomposed soap.

The breaks in the curve of Figure 3–1, at which the increases in degree of hydrolysis with increased concentration become more pronounced, are considered by Ekwall to denote points at which progressively more basic acid soaps are formed. Thus, in lieu of the initial simple NaL·HL of equation (6), he proposes that the acid soaps progress through 3NaL·HL, to a combination *averaging* 5NaL·HL at about the maximum point in the hydrolysis curve. The range of concentration for "abnormal" hydrolysis of sodium laurate (due to acid soap formation) therefore extends from 0.0075*N* to 0.05*N*. In the light of Ekwall's theory, this would correspond to the range in which the simplest (double) associated laurate ions to the most associated (probably about ten-fold) laurate ions are formed.

The effects of concentration, temperature, and composition of the soap on the degree of hydrolysis in aqueous soap solutions have been studied extensively by several investigators.[6–10] The results of studies along these lines by Powney and Jordan[6] and by McBain, Laurent, and John[7] are shown in Table 3–1 and in Figures 3–2 to 3–11. Powney and Jordan summarize the hydrolysis reactions as follows: (a) In very dilute solutions the hydrolysis is normal, and the hydrolytic fatty acid present is less than its saturation concentration; (b) the hydrolysis minimum occurs when the concentration of fatty

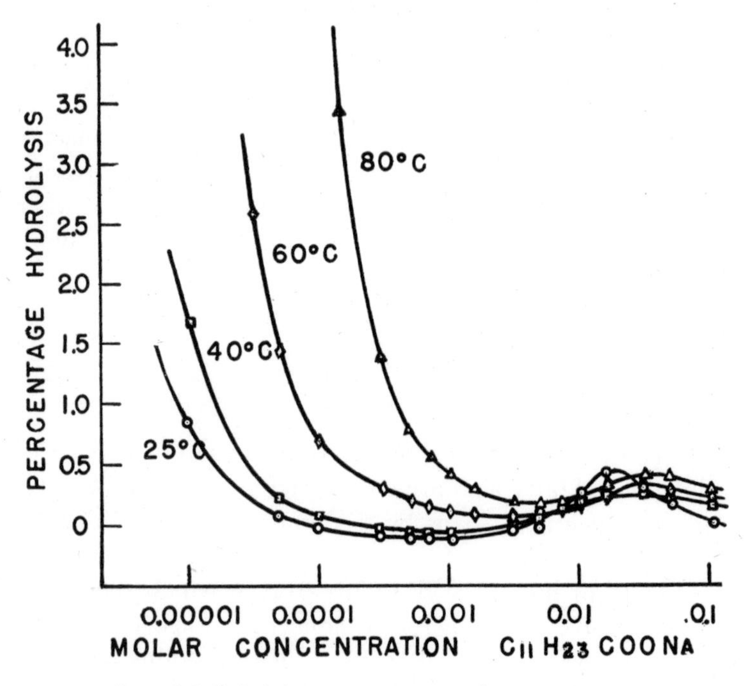

Figure 3–2. Hydrolysis-concentration curves for sodium laurate at various temperatures.[6]

acid reaches the solubility limit; (c) in regions where the hydrolysis rises to a maximum, the amount of fatty acid may be many times greater than the saturation concentration. This agrees with the previously discussed views of Ekwall. McBain takes exception to the calcu-

[6] Powney, J., and Jordan, D. O., *Trans. Faraday Soc.*, **34**, 363–71 (1938).
[7] McBain, J. W., Laurent, P., and John, L. M., *J. Am. Oil Chemists' Soc.*, **25**, 77–84 (1948).
[8] McBain, J. W., and Martin, H. E., *J. Chem. Soc.*, **105**, 957–77 (1914).
[9] Snell, F. D., *Ind. Eng. Chem.*, **24**, 76–80 (1932).
[10] Vizern and Guillot, *14me Congr. chim. ind.*, Paris, October, 1934.

lations by Powney and Jordan of the concentrations of fatty acid in the soap solutions and considers the concentrations to be actually less than the saturation concentration throughout the entire range investigated. It appears, however, that this contradiction arises not from errors in calculation but from a misunderstanding of Powney's and Jordan's assumed intent that their reference to fatty acid in excess of saturation denotes in reality fatty acid present in acid soap.

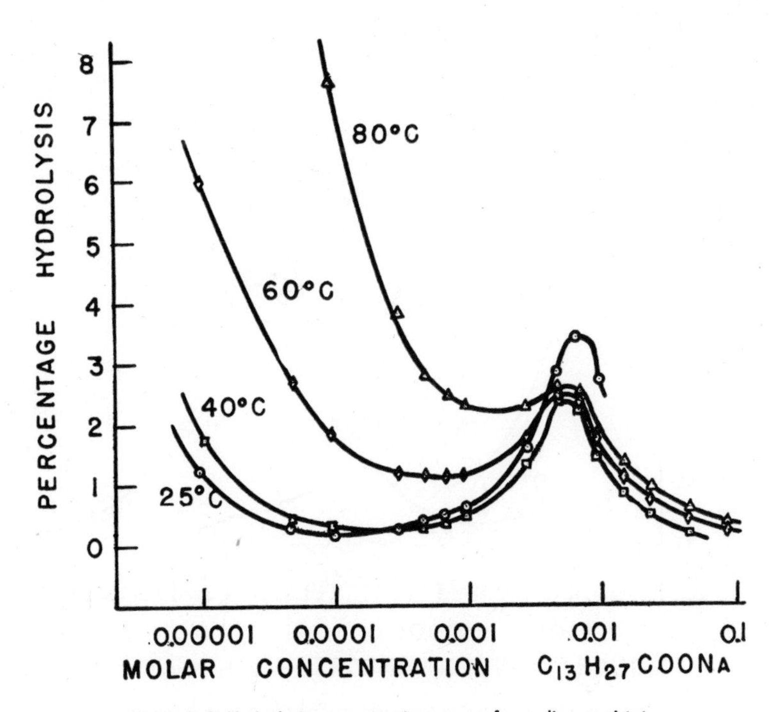

Figure 3–3. Hydrolysis-concentration curves for sodium myristate at various temperatures.[6]

For solutions of a given homologous series of soaps, the maximum in the hydrolysis-concentration curve is attained at a lower concentration the higher the molecular weight of the soap. Further, the actual extent of hydrolysis at the maximum is greater as the molecular weight of the soap increases. In comparing solutions wherein the soaps are of the same chain-length fatty acid, it is found that the greater the unsaturation of the paraffin chain the less the degree of hydrolysis. This is exemplified by the C_{18} soaps, sodium stearate, oleate, linoleate, and ricinoleate. In other words, unsaturation has the effect of "reducing" the properties of the soap to those of a lower *saturated* soap. This analogy will be noted throughout the discussions of other properties which are to follow. Finally, to conclude this summarization of the effects of soap composition on hydrolysis, the polar constituents must be considered. Thus, between sodium and potassium soaps of the fatty acid, the degree of hydrolysis of the latter is the greater.

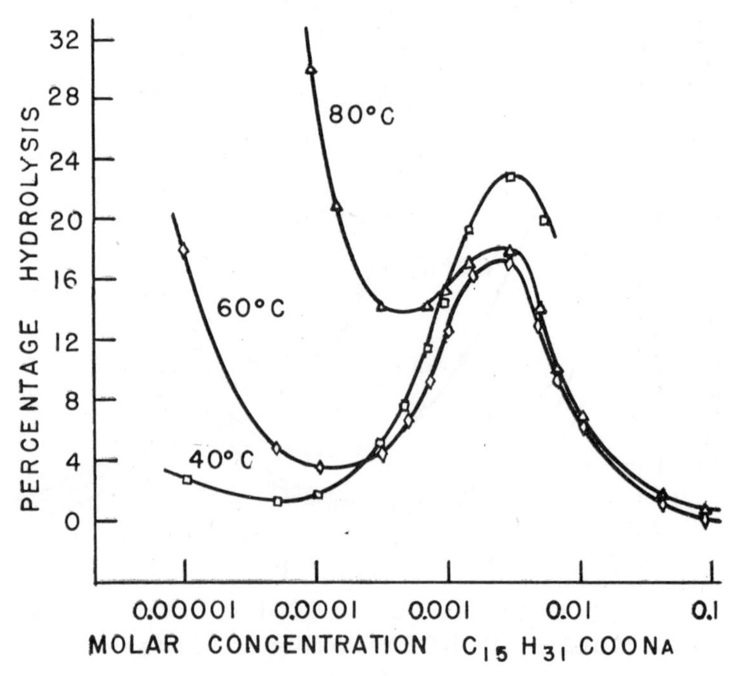

Figure 3–4. Hydrolysis-concentration curves for sodium palmitate at various temperatures.[6]

In the case of solutions of a given soap at different temperatures, it may be said that there is no consistent correlation between temperature and the degree of hydrolysis at the maximum point. At concentrations below that which shows the hydrolysis maximum, the degree of hydrolysis consistently increases with increased temperature. The fact that the concentration at which the hydrolysis maximum occurs is quite independent of the temperature is very significant in our discussion of detergency in that it correlates well with a similar independence of some surface-activity properties, as will be discussed in Chapter 4. Also of significance is the fact that the concentrations for the hydrolysis maxima of the various soaps are in the range of soap concentration used in laundering.

The pronounced difference in character between the two ends of a detergent molecule accounts for much of its behavior in water. The hydrocarbon portion of the molecule becomes less soluble in water as a homologous series is ascended or, conversely, becomes more soluble with increased unsaturation. Thus, a salt such as sodium acetate might be considered analogous to soap, except that the acetate por-

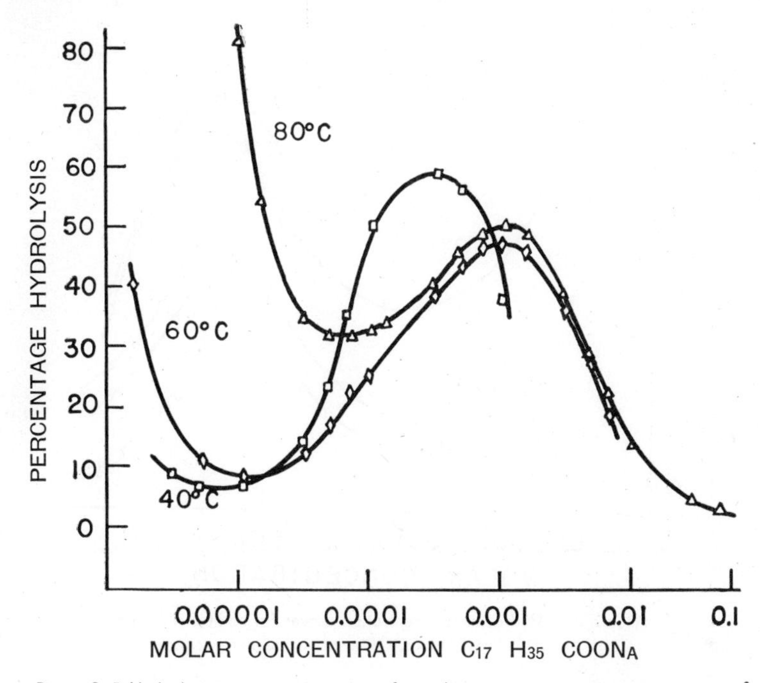

Figure 3–5. Hydrolysis-concentration curves for sodium stearate at various temperatures.[6]

tion of the molecule or, more properly, the acetate ion is much more soluble in water than a laurate, myristate, palmitate, stearate, or other fatty ion of a soap. In opposition to the low solubility of the hydrocarbon group of a detergent, we have the high solubility of such polar groups as $—COONa$, $—COO^-$, $—COSO_2ONa$, $—COSO_2O^-$, $—COSO_2Na$, and $—COSO_2^-$. Thus, we have any of these latter groups tending to pull the hydrocarbon group into solution and the hydrocarbon group opposing solution. There is much indirect evidence to indicate that the solubilities of the various polar groups listed are quite similar; consequently, we then have decreasing over-all solubility as we ascend a hydrocarbon series, more or less independent of the nature of the polar group. Likewise, in the case of soaps, the "acidity" of the fatty acids becomes weaker as we ascend a series, from which we would expect the corresponding increase in extent of hydrolysis already discussed. In the case of anionic synthetic detergents, any influence of the hydrocarbon group on acidity

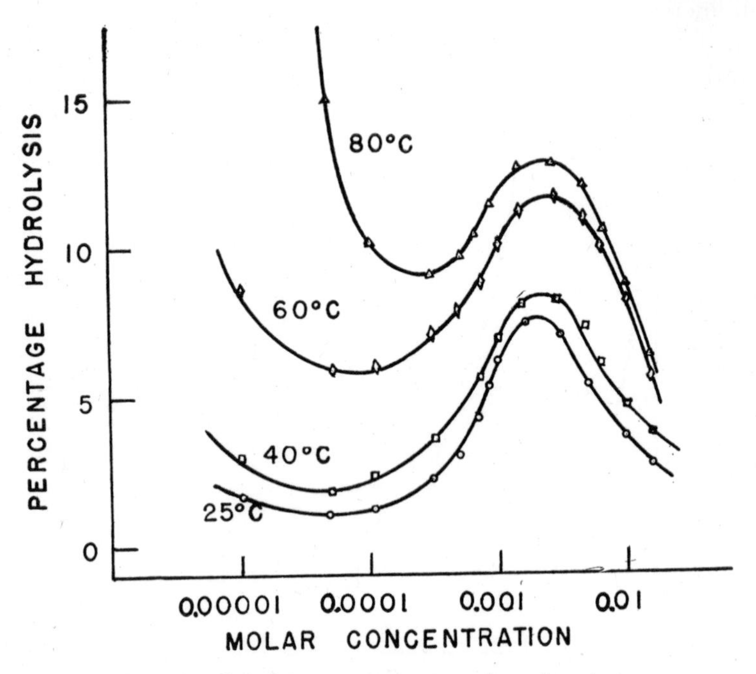

Figure 3–6. Hydrolysis-concentration curves for sodium oleate, $CH_3(CH_2)_7CH : CH(CH_2)_7COONa$, at various temperatures.[6]

is weak at most and over-shadowed by the strong influence imparted by sulfate or sulfonate groups.

Our discussion leads now to consideration of the hydrolysis of synthetic detergents. These compounds in general differ from soaps primarily in difference in their susceptibility to hydrolysis. This has been one of the prime motivating factors in their development. Many synthetic detergents, particularly the nonionic and cationic compounds, while being excellent for certain special applications, are not readily suitable for use in laundering and will not be included in the present discussion. On the other hand, synthetic detergents of other types, which are at least *approaching* soap in adaptability to commercial laundry use, will be considered because of the imminent possibility that their further development will place them in a more favorable position.

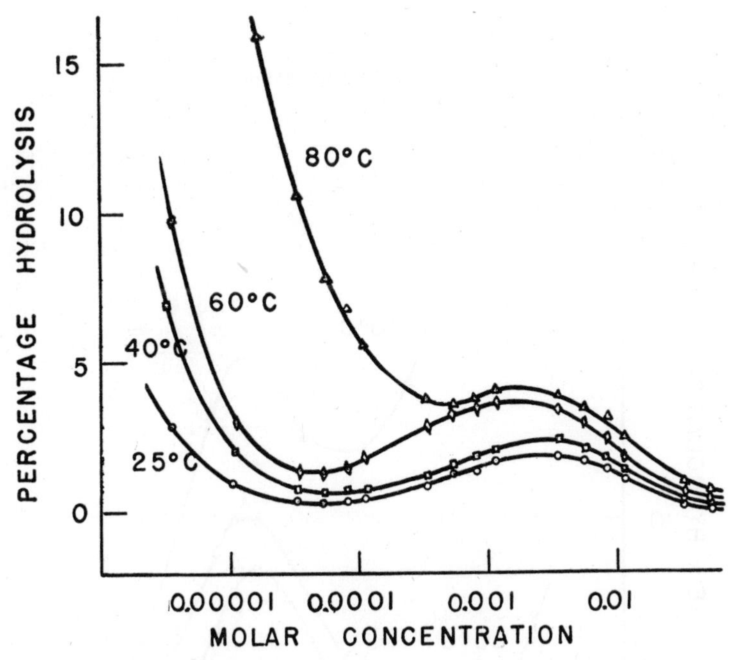

Figure 3–7. Hydrolysis-concentration curves for sodium linoleate, $CH_3(CH_2)_5CH : CHCH : CH(CH_2)_7COONa$, at various temperatures.[6]

The more adaptable synthetic detergents are sulfates and sulfonates of six types, as represented by the following examples, all of which are so-called anionic materials:

(a) Alkyl sulfates—

$CH_3(CH_2)_{11}OSO_2O\vdots Na$ (sodium dodecyl sulfate);

(b) Alkyl sulfonates—

$CH_3(CH_2)_{11}SO_3\vdots Na$ (sodium dodecyl sulfonate);

(c) Esters of polyhydric alcohol-sulfonates—

$CH_2OCO(CH_2)_{10}CH_3$
|
$CHOH$
|
$CH_2OSO_2O\vdots Na$ (sodium glyceryl monolaurate sulfate);

(d) Alkylaryl sulfonates—

$CH_2(CH_2)_{10}CH_3$ (on a benzene ring, with SO_3 | Na para)

SO_3 | Na (sodium lauryl benzenesulfonate);

(e) Sulfated or sulfonated alkyl amides—

$C_{17}H_{33}CON(CH_3)CH_2CH_2SO_3$ | Na (sodium, 1, (NN′ oleyl methyl) aminoethane sulfonate);

(f) Sulfated or sulfonated esters—

$CH_3(CH_2)_{10}COOCH_2SO_3$ | Na (sodium methyl laurate sulfonate).

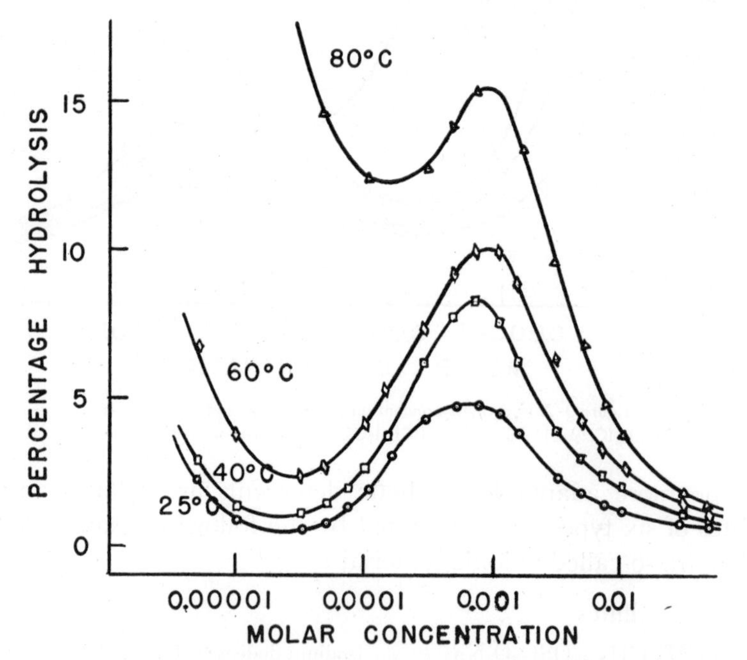

Figure 3–8. Hydrolysis-concentration curves for sodium ricinoleate, $CH_3(CH_2)_5CHOHCH_2CH : CH(CH_2)_7COONa$, at various temperatures.[6]

These compounds have the point in common with each other, and in distinction from soaps, of being salts of relatively strong acids. Thus, though they are all anionic detergents like soap because of dissociation (where indicated in the above formulas by brackets) to give sodium cation and complex sulfate or sulfonate anion, subsequent association of their anions with hydrogen ion is not nearly so pronounced as in the case of the soap anions. Therefore, the degree of hydrolysis to form an organo-sulfuric or sulfonic acid and the pH in solutions of these compounds will be considerably less than that in solutions of soaps of corresponding fatty acids.

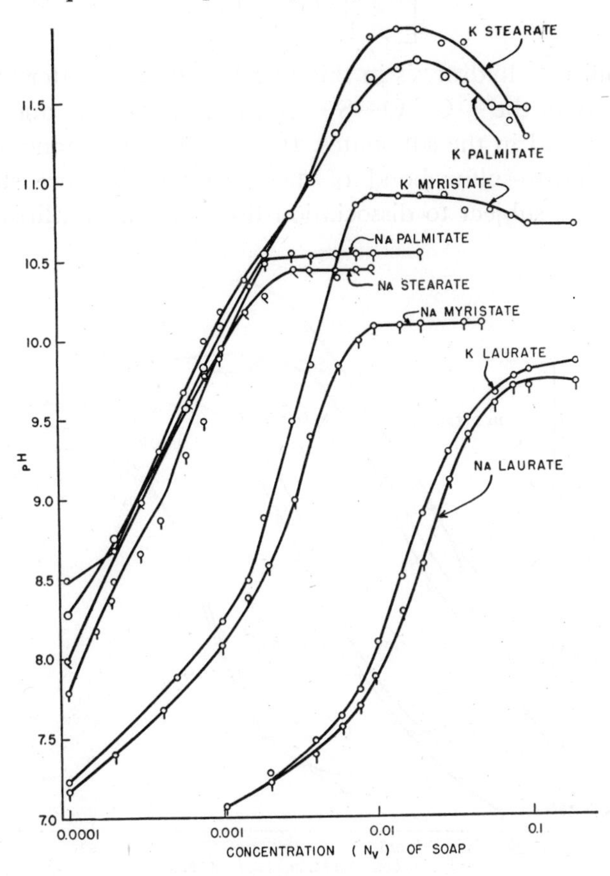

Figure 3–9. Comparison of the pH of solutions of sodium and potassium soaps at 25°C.[7]

In the case of the above-listed compounds, hydrolysis at another common point in the molecule must be considered, namely, at the linkage between the sulfate or sulfonate group and the organic portion of the molecule. The sulfate linkage may be represented as:

$$\left[\begin{array}{ccccccccc} & & H & & & & O & & \\ & & | & & & & | & & \\ R & - & C & - & O & - & S & - & O \\ & & | & & & & | & & \\ & & H & & & & O & & \end{array}\right]^{-} Na^{+}$$

and the sulfonate linkage as:

$$\left[\begin{array}{ccccccc} & & \mathrm{H} & & \mathrm{O} & & \\ & & | & & | & & \\ \mathrm{R} & — & \mathrm{C} & — & \mathrm{S} & — & \mathrm{O} \\ & & | & & | & & \\ & & \mathrm{H} & & \mathrm{O} & & \end{array}\right]^{-} \mathrm{Na}^{+}$$

Susceptibility to hydrolysis in this case becomes a matter of the relative stability of the —C—O—S— bond in the sulfates and the direct —C—S— bond in the sulfonates. It may be said in general that the direct carbon-to-sulfur bond of the sulfonates is more stable and, therefore, less subject to dissociation than the oxygen linkage of the

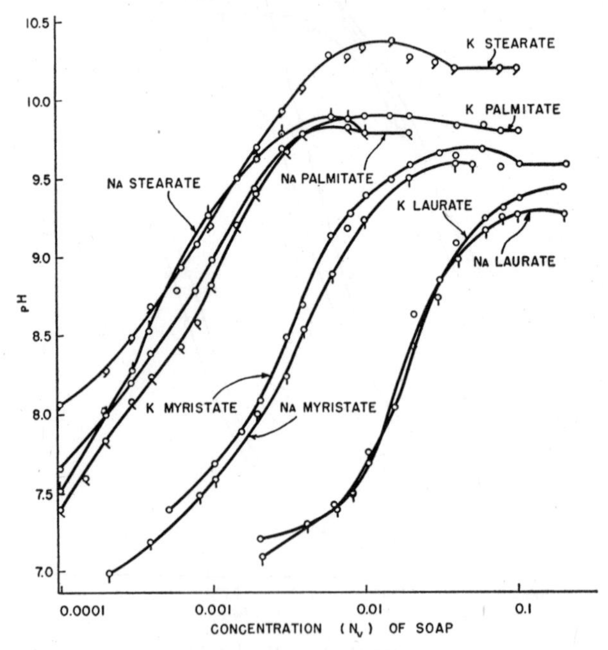

Figure 3–10. Comparison of the pH of solutions of sodium and potassium soaps at 50°C.[7]

sulfates. It follows then that, in the case of any two synthetic detergents which are alike in composition except that one is a sulfate and one a sulfonate, the sulfonate will be the more stable of the two. However, it should be pointed out that such a comparison of stability does not necessarily parallel relative detergent efficiency.

One further significant point of dissociation to be considered is that of the —O— ester linkage of the ester-type detergents represented by examples (c) and (f), above. This linkage limits the stability of such compounds in the presence of strong acids or high alkalinity, but permits reasonable stability in the presence of moderate alkalinity.

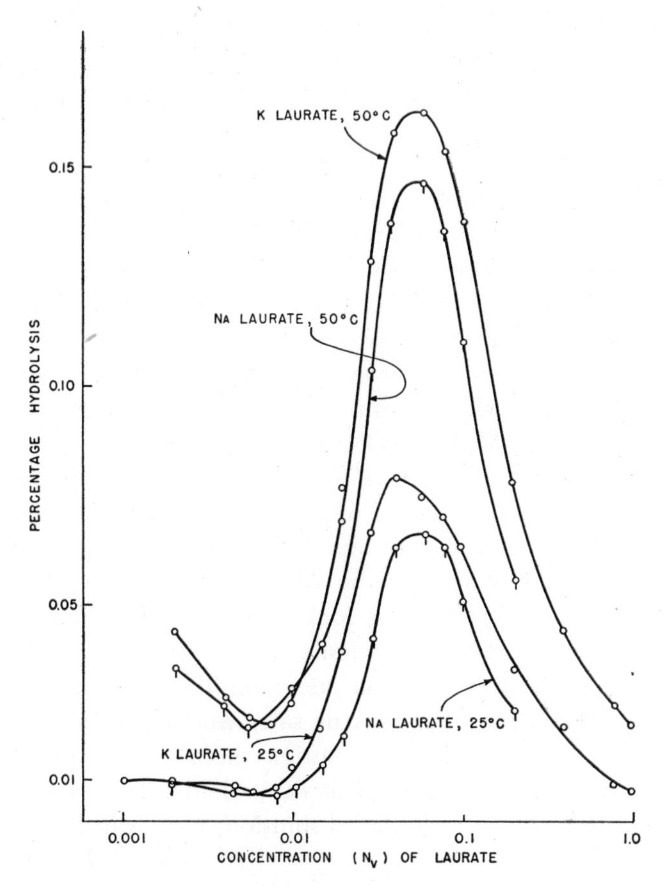

Figure 3–11. Percentage hydrolysis-concentration curves for sodium and potassium laurates.[7]

Nature of Detergents in Solution

The discussion heretofore in this chapter has been without reference to the physical states in which soap and its products of hydrolysis are present in aqueous solution. Krafft and coworkers[11-13] were perhaps

[11] Krafft, F., and Wiglow, H., *Ber.*, **28**, 2573–82 (1895).
[12] Krafft, F., and Strutz, A., *Ber.*, **29**, 1328–31 (1896).
[13] Krafft, F., *Ber.*, **32**, 1584–96 (1899).

the first to recognize the colloidal characteristics of such solutions. Thus, in attempting to determine the molecular weights of soaps from elevation of the boiling point of water, they found at low concentrations that the boiling point was raised to some extent but that higher concentrations caused the boiling point to recede. From the fact that soap in the higher concentrations does not materially affect the boiling point of water they concluded that soap in aqueous solution is a colloid.

In 1910, McBain and Taylor[14] reported their preliminary study of the high electrical conductivity of soap solutions, from which it was necessary to conclude that soap in aqueous solutions was not present as neutral colloid, as previously considered. This paper was followed by another[15] in which they reported the conductivities of aqueous sodium palmitate solutions between 0.01*N* and 2*N* at 90°C as being about one-half as great as those of equivalent sodium acetate solutions. Similar work with sodium myristate and sodium laurate was reported in 1913 by McBain, Cornish, and Bowden.[16]

This preliminary work was followed by a series of papers by McBain and coworkers[17-26] in which a theory of the existence of colloidal electrolytes was developed. McBain proposes a new and extensive "state of matter"—the colloidal electrolytes, which are salts in which one form of the ions (in the case of soap, the fatty anions) has been replaced by ionic micelles. Starting with dilute solutions and with increasing concentration, there is a gradual transition from ordinary electrolytes (true salts) and simple ions to typical colloidal electrolytes. In dilute solution there are present both undissociated and dissociated soap as crystalloids of simple molecular weight; in sufficiently concentrated solution there is little else present

[14] McBain, J. W., and Taylor, M., *Ber.*, **43**, 321–2 (1910).
[15] McBain, J. W., and Taylor, M., *Z. phyzik. Chem.*, **76**, 179–209 (1911).
[16] McBain, J. W., Cornish, E. C. V., and Bowden, R. C., *J. Chem. Soc.*, **101**, 2042–56 (1912).
[17] McBain, J. W., *J. Soc. Chem. Ind.*, **37**, 249–52 (1918).
[18] McBain, J. W., and Salmon, C. S., *J. Am. Chem. Soc.*, **42**, 426–60 (1920).
[19] McBain, J. W., Taylor, M., and Laing, M. E., *J. Chem. Soc.*, **121**, 621–33 (1922).
[20] McBain, J. W., and Jenkins, W. J., *J. Chem. Soc.*, **121**, 2325–44 (1922).
[21] McBain, J. W., and Bowden, R. C., *J. Chem. Soc.*, **123**, 2417–30 (1923).
[22] McBain, J. W., *Am. Dyestuff Reptr.*, **12**, 822–6 (1923).
[23] McBain, J. W., *J. Am. Chem. Soc.*, **50**, 1636–40 (1928).
[24] McBain, J. W., *Brit. Assoc. Advancement Sci.*, Rept. 88th meeting, 2–31 (1930).
[25] McBain, J. W., and Liu, T. H., *J. Am. Chem. Soc.*, **53**, 59–74 (1931).
[26] McBain, J. W., and Bolduan, O. E. A., *J. Phys. Chem.*, **47**, 94–103 (1943).

than colloid plus simple cations. As the concentration increases there is also a continuous transition from typical colloidal electrolyte (where the micelles are highly charged), through slightly charged colloids, to typical neutral colloids; however, concentrations where neutral colloid may prevail are not attained in laundering.

The characteristics of soap solutions with respect to physical states have been summarized by McBain as follows:

a. Ionic micelles have high electrical conductance because of their multiple charge and the fact that, with their polyvalence, they move faster in an electric field than do simple ions; for example, the conductance of one ionic micelle containing, say, 10 palmitate ions has been reported to be 20–30 times greater than that of the single palmitate ion.

b. Upon sufficient dilution of soap solutions, the colloid breaks up into simple ions.

c. At about 0.01*N* (in the approximate range for laundering) there are present in solution alkali cations, ionic micelles, simple fatty anions, undissociated neutral colloid, and undissociated simple molecules.

d. Formation of colloidal electrolyte is favored by lowering of temperature, increase in molecular weight, presence of more than one soap, and addition of crystalloidal electrolyte (see Chapter 8).

The theory of the existence of ionic micelles in soap solutions has not gone unchallenged. Linderstrom-Lang[27] has formulated a theory for the interaction between chain molecules and chain ions and indicates as probable that the cohesive forces between the chain ions in soap solutions are able to occasion the slight osmotic pressure, high viscosity, gel formation, and other unusual properties of such solutions. He considers that soaps in aqueous solution at 90°C are molecularly dispersed for a considerable concentration interval and that the intertwining of the fatty chains through the body of the solution determines its properties. McBain[23] presents a refutation of Linderstrom-Lang's theory, based on ultrafiltration, migration, viscosity, and hydrolysis data. First he shows that, where quantitative data for activity and conductivity indicate the presence of crystalloid only, the whole of the soap solution passes through dense ultrafilters, but

[27] Linderstrom-Lang, K., *Compt. rend. trav. lab. Carlsberg*, **16,** No. 6, 1–47 (1926).

that, where such data indicate complete formation of colloid, the whole of the soap may be filtered off by only moderately dense ultrafilters. Further, in the case of migration, he shows that, in solutions containing colloid, sodium and potassium are moving to the anode, in the opposite direction from that of true Na^+ and K^+ ions, due to being carried in undissociated colloid. Next, the enormous effect on viscosity of soap solutions, as produced by change in temperature or addition of small amounts of electrolyte, cannot be explained on the basis of Linderstrom-Lang's theory of a "tangle" of fatty ions. Finally, McBain points out that, if soaps acted in solution as simple electrolytes, their hydrolysis alkalinity would increase steadily with increased concentration, instead of first increasing, passing through a maximum, and diminishing again in more concentrated solution as the hydrolyzable simple fatty ions are replaced by ionic micelles.

Harkins, Mattoon, and Corrin[28] point out the relative solubilities of long paraffin chains (insoluble) and of the corresponding fatty acids (practically insoluble) and soaps in water. On the basis that potassium laurate gives a clear solution at concentrations even somewhat above 35 per cent and potassium myristate above 26 per cent, they consider that the spontaneous aggregation of the soap into micelles to give such high solubilities is due to the greater reduction of free energy than is the case when going into solution as single molecules. The extent of the free energy reduction possible from such aggregation will be evident from discussions of surface activity which follow in Chapter 4.

The presence of micelles in any but very dilute aqueous detergent solutions is so well established as to be no longer challenged. On the other hand, the picture of the structure of the micelles is by no means so clear. The basic argument in this respect revolves around spherical *vs.* lamellar micelles. Many of the studies of the problem have been based on conductivity and other measurements which follow in at least a qualitative way the changes in composition of the solutions with changes in detergent concentration. Such changes have extended from very dilute solutions on the one hand to quite high concentrations entirely beyond the range applicable in ordinary laundry washing operations. A discussion of the complete range, although of great

---

[28] Harkins, W. D., Mattoon, R. W., and Corrin, M. L., *J. Am. Chem. Soc.*, **68**, 220–8 (1946).

general technical interest, is beyond the scope of this dissertation; therefore, our comments will be confined for the most part to the lower detergent concentrations. Where transitions with change in concentration are involved, the direction of change in concentration from low to high will be implied.

Hartley[29] points out that, from the standpoint of molecular forces, micelle formation in solutions of paraffin-chain salts might more properly be considered as an aggregation of *water* than of solute. Thus, it is the strong interattraction between water molecules tending to exclude hydrophobic detergent species from solution that plays a greater part in micelle formation than attraction of the detergent species for each other. It is this difference which accounts for the pronounced free surface energy of the solute.

As between detergent anion and molecule, the exclusion of the anion from solution may be expected to precede that of the molecule, because of the lesser "solubility" of the polar end of the anion. This, together with the significant dissociation of detergent molecules in dilute aqueous solution, substantiates the generally accepted view originally advanced by McBain that the initial micelle formation constitutes an aggregation of anions. Further, the smaller the anion, the smaller will be the anionic micelle and, consequently, the smaller will be the reduction of free surface energy brought about by the aggregation. This may very well account for the fact that the lower the molecular weight of the aggregating species the higher is the concentration necessary to effect the aggregation.

McBain proposes, in addition to the small, highly-charged ionic micelles in rather dilute solutions, the presence of lamellar highly organized aggregates of soap molecules in more concentrated solutions. These latter aggregates are the micelles referred to by McBain as "neutral" in his earlier work but which he now recognizes as being weakly conducting.[30] Unfortunately, the literature is vague as to the detergent concentrations at which the transition from ionic to lamellar micelles takes place, if such a transition actually does occur. However, it is quite certain that the highly organized lamellar micelles (as will be discussed below) do not exist to an appreciable extent in unbuilt

[29] Hartley, G. S., "Aqueous Solutions of Paraffin-Chain Salts," Paris, Hermann & Cie, 1936.

[30] McBain, J. W., "Solubilization and Other Factors in Detergent Action," "Advances in Colloid Science," pp. 99–142, New York, Interscience Publishers, Inc., 1942.

solutions at laundry concentrations. Certainly there has been no direct evidence, as from x-ray diffraction studies, of organized micellar structure at such low concentrations.

Hartley[29] presents a strong case for spherical ionic micelles, the schematic representation of which is shown in Figure 3–12. His reasoning for this structure is predicated on the fact that its roughly spherical surface represents a reduction to the minimum possible area of the water-detergent interface and, thus, a reduction to the minimum possible total free surface energy. He particularly stresses that there is no justification for a highly organized structure in the micelle, other than the generally selective orientation of the individual fatty ions (discussion of this selective orientation is reserved for Chapter 4). Whether or not this unorganized structure continues into higher solution concentrations is not a question pertinent to the present discussion; however, at the low concentrations with which we are interested, the spherical micelle pictured by Hartley might well be synonymous with McBain's small highly-charged ionic micelle.

The structure of detergent micelles at higher solution concentrations (5 per cent and above) has been studied extensively by x-ray techniques, particularly in connection with the action of solubilization. The x-ray diffraction patterns thus produced demonstrate quite conclusively the presence of a highly organized structure, perhaps somewhat as illustrated schematically in Figure 3–13, taken from the work of McBain.[30] Our immediate interest in such micelles is limited to such special applications in laundering as "spotting."

Hess, Kiessig, and Philippoff[31-33] postulate that lamellar soap micelles are built up of sheets of oriented molecules with —COONa groups forming one surface of the sheet and the terminal —CH_3 groups forming the other surface. In the micelles the sheets are so arranged that pairs of —COONa surfaces face each other, as do also pairs of —CH_3 surfaces. Such arrangement is illustrated in Figure 3–13. This latticelike arrangement of the soap molecules in the micelle is not identical with that in solid soap crystals, an essential difference being that water adds in the micelles between the —COONa groups in the direction of the length of the molecules. The micelles

---

[31] Hess, K., *Fette u. Seifen*, **46**, 572–5 (1939).
[32] Hess, K., Kiessig, H., and Philippoff, W., *Fette u. Seifen*, **48**, 377–84 (1941).
[33] Kiessig, H., and Philippoff, W., *Naturwissenschaften*, **27**, 593–5 (1939).

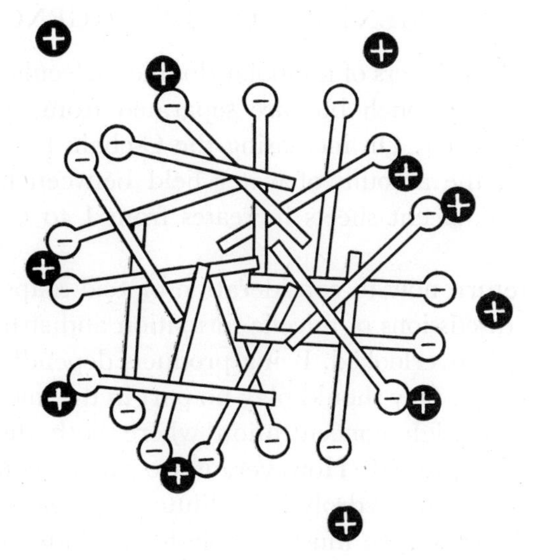

Figure 3–12. Schematic representation of a spherical micelle.[29]

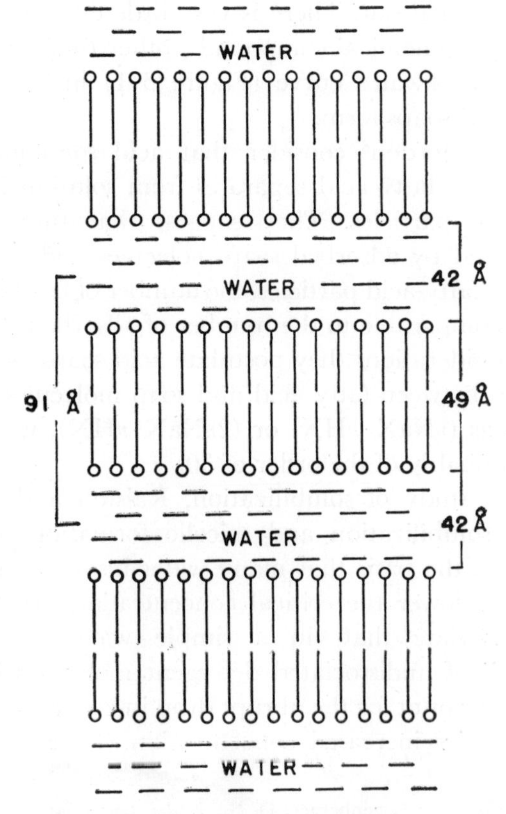

Figure 3–13. Highly idealized representation of the lamellar **structure of a micelle in a 9.12 wt.% sodium oleate solution.** (After McBain[30])

exist accordingly as layers of nonpolar double molecules which, unlike solid soaps, do not touch but are separated from one another by layers of added water. On increasing the C-chain length of the soap from C_4 to C_{10}, the amount of water held between the —COONa surfaces of the adjacent sheets increases from 1 to 6 molecules per mol of soap.

We must return now to consideration of acid soaps which, in the past extensive discussions of micelle formation and structure, seem to have been almost overlooked. Being products depending on dissociation and hydrolysis, they should play no part in the micelles present in soap solutions of high concentration where both dissociation and hydrolysis are suppressed. However, if we are to consider pictures such as Ekwall's of the hydrolysis in dilute soap solutions to be even approximately correct, we must be able to correlate acid soaps with the ionic micelles also known to be present.

The uncertainty of the stoichiometry of acid soap formation has already been pointed out. There is very little evidence in support of acid soaps as true chemical compounds, other than such evidence as the "breaks" in Ekwall's curve (Figure 3–1) in the concentration range where acid soaps form.

Powney and Jordon[6] consider that acid soap formation commences as soon as fatty acid separates from solution in the form of ultramicroscopic particles. They suggest that these particles are initially stabilized by adsorbed soap molecules and that, due to the smallness of the fatty-acid particles, the number of fatty-acid molecules may be quite comparable to the number of adsorbed soap molecules. From these considerations they postulate acid soaps as being adsorption complexes between fatty acid and soap molecules, expressed by formulas such as (nNaX · nHX) or (2nNaX · nHX), where the lowest value of n is probably of the order of 50.

In a recent study of solubilization, Ralston and Eggenberger[34] consider that solubilization and micelle formation may be allied phenomena. On the basis that many water-insoluble nonelectrolytes can appreciably lower the critical concentration for micelle formation, they postulate that, in a simple water-detergent system, "solubilization" of undissociated detergent molecules into the ionic micelles might account for the abrupt drop in equivalent conductivity which accompanies increased concentration. Thus, a reaction such

---

[34] Ralston, A. W., and Eggenberger, D. N., *J. Am. Chem. Soc.*, 70, 983–7 (1948).

as the following to produce mixed ionic-molecular micelles might be considered analogous to the solubilization action which will be discussed in Chapter 5:

$$\underset{\substack{\text{(soap}\\ \text{molecules)}}}{x\text{Na}\cdot\text{S}} + \underset{\substack{\text{(ionic}\\ \text{micelles)}}}{y(\text{S}^-)} \rightarrow \underset{\substack{\text{(ionic-molecular}\\ \text{micelle)}}}{(\text{Na}\cdot\text{S})_x(\text{S}^-)_y} \tag{7}$$

The essential difference between the ionic and the ionic-molecular micelles is that the latter has solubilized the undissociated soap molecules.

Ralston and Eggenberger propose that their ionic-molecular micelle is lamellar, a structure we have chosen to consider to be nonexistent at low concentrations. Further, they give no consideration to the fatty acid. However, without entering into the argument of micelle structure, it is considered that a reasonable picture of "acid soap" formation can be developed on the basis of solubilization of fatty acid in ionic micelles. Powney and Jordan have pointed out that, in going from very dilute to more concentrated soap solutions, there is a short range of visible turbidity just above the limit of solubility of fatty acid. This is also in the range of minimum micelle formation. The subsequent clearing of such solutions upon further increase in concentration may well be associated with greater formation of micelles and, consequently, solubilization or even solution of the fatty acid in these micelles.

Summarization

Summarization of the foregoing discussion will be devoted primarily to the general range of detergent concentrations used in laundering. This range is of the order of 0.1 to 0.3 per cent, the actual concentration used in a specific case depending on such factors as composition of the detergent, temperature at which washing is to be conducted, and "heaviness" of soil. The laundryman's gauge for the proper soap concentration is the depth of "running suds."

Of particular significance is the fact that the detergent concentrations used in actual practice to attain maximum detergency correspond quite closely with the critical concentrations for formation of micelles. This correlation will be discussed at length in later chapters; for the present it is sufficient to say that the factors which determine detergent power and the factors which determine micelle formation are quite synonymous. Therefore, in this summarization we may say that detergent concentrations used in laundering correspond approximately to critical micelle concentrations. One component of the detergent solution—the ionic micelle—is then immediately established.

Aqueous soap solutions at laundry concentrations probably are composed of

(a) Fatty anions,

(b) Metallic (sodium) cations,

(c) Undissociated soap molecules,

(d) Ionic soap micelles in probably the *incipient* state of formation, that is, in a relatively low degree of aggregation, and

(e) Acid soap "aggregates," with the proviso that acid soap may be, rather than a separate entity, actually the first manifestation of micelle formation.

The equilibrium concentrations of these various components are established by these factors: soap composition and temperature of the system.

The effects of change in concentration on the identities of the components present in soap solutions perhaps are best illustrated schematically as follows:

(a) At high dilution—

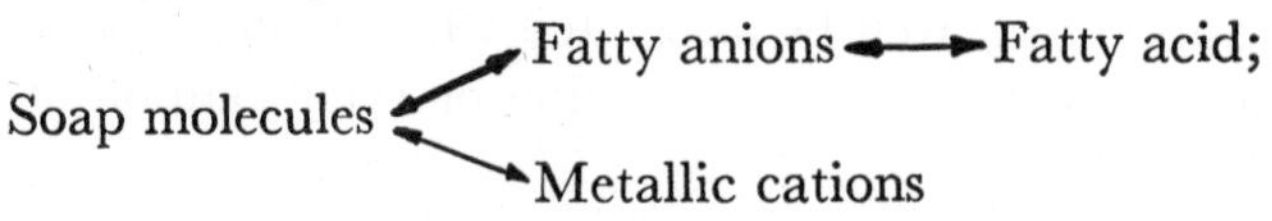

(b) At concentrations where the fatty acid formed just exceeds the solubility limit—

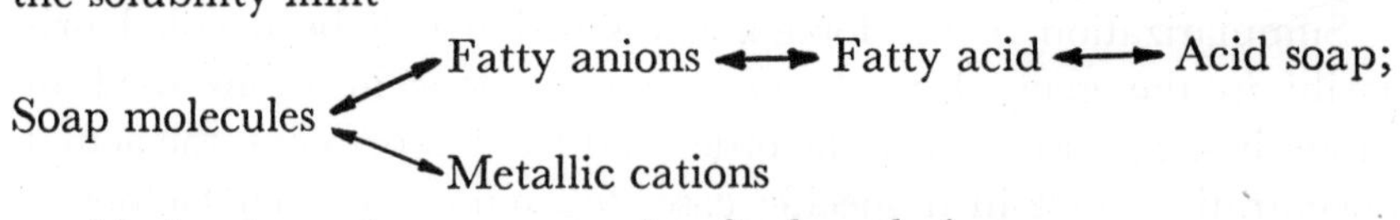

(c) At about the concentration for laundering—

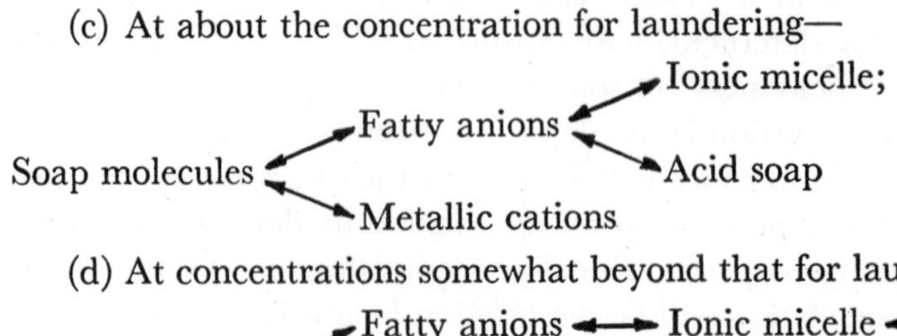

(d) At concentrations somewhat beyond that for laundering—

Fatty anions ⟷ Ionic micelle ⟷ Ion-molecule micelles;
Soap molecules
Metallic cations

(e) At high concentrations (perhaps 5 per cent and above)—

Soap molecules ⟷ Neutral micelles.

Briefly, higher dilution of the solution favors dissociation and hydrolysis, to the exclusion of aggregation; lower dilution (higher concentration) favors aggregation, to the exclusion of dissociation and hydrolysis.

What has been summarized above for soap solutions holds equally well for solutions of anionic synthetic detergents, except that the hydrolysis products—fatty acid and acid soap—must be omitted. Cationic detergents involve fatty cations rather than fatty anions. Solutions of nonionic detergents should contain only molecules below the critical micelle concentration; above this concentration, the conditions of (e) should obtain.

As will be more evident from later discussions, probably the most important field for future studies of the nature of detergent solutions lies in determinations of the relative amounts of various components present in solution under given sets of conditions of composition, over-all concentration, and temperature.

4

Surface Activity of Unbuilt Detergent Solutions

Surface Tension

The attraction between molecules within the body of a liquid such as water is distributed equally in all directions, but at the surface this attraction results in an unbalanced force tending to pull the surface molecules into the body and to adjust the surface to a minimum area, as illustrated in Figure 4–1. Any attempt to increase the surface area of such a liquid, as by the formation of droplets or films, is opposed by this force, the extent of which is defined as the *surface tension*, or the force per unit length on the surface of a liquid which opposes expansion of the surface area. The work necessary to increase the area of a liquid surface is equal to the *surface energy* and, per unit area, is numerically equal to the surface tension. In a sense, surface energy may be considered as the equal antithesis of surface tension. Due to their numerical equality when expressed in energy units, the two terms are frequently considered synonymous.

Water is an unusual substance in that it has considerably the highest surface tension of most liquids of comparable molecular weight. This high surface tension, with resulting relatively poor wetting ability when compared to that of many organic liquids, is due to the high attractive forces and association between the water molecules and, thus, to its high free surface energy.

Starting with a solvent such as water, having a definite surface tension, the addition of many kinds of solute molecules will impart to the resulting solution a lower or a higher surface tension than that of the pure solvent, depending on whether the concentration of solute molecules in the surface is greater or less than that in the body of the solution. Solute molecules which are more concentrated in the surface than in the body of a solution are said to be *positively* adsorbed.

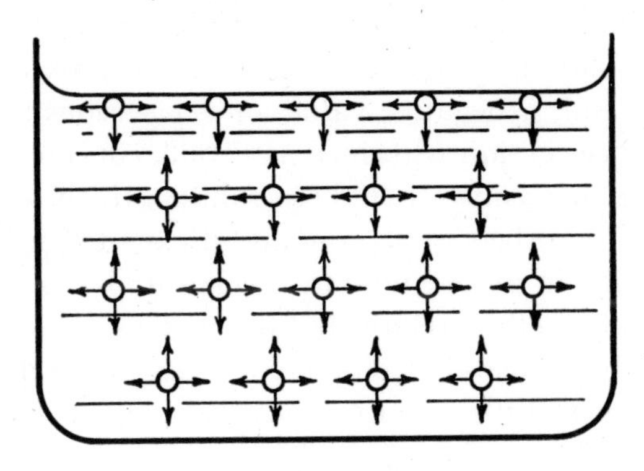

A — IN BULK

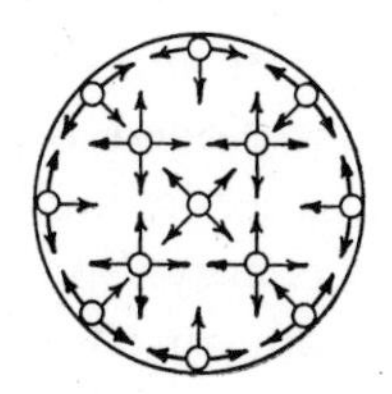

B — SPHERICAL DROPLET

(MINIMUM SURFACE AREA FOR AN UNSUPPORTED LIQUID)

Figure 4–1. Illustration of the relative fields of attraction between liquid molecules on the surface and within the bulk of the liquid.

Solute molecules which are more concentrated in the body than in the surface of the solution are said to be *negatively* adsorbed.

As determined mathematically by Gibbs,[1] a substance may become more concentrated in the surface than in the body of a solution whenever its presence in the surface decreases either its own free surface energy or the free surface energy of the solvent. Although more accurately based on *activities*, a commonly used form of the Gibbs adsorption equation is

$$\Gamma = -\frac{c}{RT} \cdot \frac{d\gamma}{dc} \tag{1}$$

---

[1] Gibbs, J. W., *Trans. Conn. Acad.*, 3, 439 (1876).

where Γ is the difference between surface and bulk *concentrations* of solute and $d\gamma/dc$ is the change in surface tension with change in bulk solute concentration. The use of concentrations rather than activities limits this equation to dilute solutions.

Examination of the Gibbs equation shows that when Γ is positive (surface concentration greater than bulk concentration) the surface tension of the solution decreases with increased bulk concentration. On the other hand, a negative value for Γ occasions a concomitant increase in surface tension with increased bulk concentration. It is of interest to note in passing that those solutes which are capable of being positively adsorbed to a pronounced extent may effect a pronounced reduction in surface tension, whereas those solutes which are negatively adsorbed effect relatively little increase in surface tension. This latter condition may be considered to be due to the bulk of the solution having very little effect on the properties of the surface.

Positive adsorption of a solute in the surface of a solution may be a manifestation that the solute is "seeking" a lower level of free surface energy. Within the bulk of an aqueous solution there may exist at the interface between water and some solute molecules or ions a high free surface energy, occasioned by inter-attraction between water molecules. The less "water-loving" the solute may be the greater will be this free surface energy. Spontaneous expulsion of such solutes to the surface results from an effort by the system to reduce this free energy. On the other hand, positive adsorption of solute also may be a manifestation of a spontaneous "effort" on the part of the solvent to reduce the free energy at its external surface. Expulsion of solute to the surface results, in effect, in a new surface wherein the interattraction between molecules is less.

Harkins, Davies, and Clark[2] propose a general criterion for adsorption from solutions. If the solvent is polar, as is water, then solutes generally will be positively adsorbed in the surface if they are less polar than the solvent, and the least polar end of the molecule will be oriented outward. Solutes more polar than the solvent are then negatively adsorbed in the surface. The explanation of the orientation of molecules at the surface of liquids was developed independently by

---

[2] Harkins, W. D., Davies, E. C. H., and Clark, G. L., *J. Am. Chem. Soc.*, **39**, 541–96 (1917).

Harkins and coworkers[2, 3] and by Langmuir.[4, 5] Starting first with a film of oil on the surface of water, they show that any active or polar group at the end of a hydrocarbon chain in the oil will tend to be drawn into the water, leaving the insoluble hydrocarbon end extending outward from the water surface. Such active or polar groups might be carboxyl groups, hydroxyl groups, double bonds of unsaturation, or the —COONa group of a sodium soap. Polar groups will "drag" into water a short hydrocarbon chain, such as acetate, to give true solution with very little selective adsorption in the surface. As the length of the chain increases, the solubility of the compound with one end polar and the other end nonpolar rapidly decreases, with corresponding increase in positive adsorption.

This leads then to the explanation of surface behavior in the specific case of aqueous detergent solutions. Here we have the case of a group of compounds whose molecules or fatty ions have contrasting hydrophilic and hydrophobic ends, the relative strengths of "hydrophilicity" and "hydrophobicity" of which are such as to impart pronounced positive adsorption. In an aqueous detergent solution, the hydrophobic portion may be considered to be continually attempting to "pull" the detergent out of solution, in order to reduce its total free surface and, thus, its total free surface energy. Such attempt is best rewarded at the surface of the solution where the detergent molecules or fatty ions become oriented, somewhat as shown in Figure 4–2, and where the hydrophobic portion in effect forms a new oily film phase wherein the characteristic of insolubility in the water is most fully satisfied. Not only is the detergent oriented in the solution surface in this manner, but, for the same reason, it becomes more concentrated in the surface than in the body of the solution at normal over-all concentrations.

As previously pointed out, this concentration in the surface particularly favors the lowering of the surface tension of the solution. More specifically, in the case of a detergent solution, there are fewer water molecules in the surface to be attracted by water molecules in the interior, and the total of the attractions between detergent molecules and water molecules in the surface is less than that which existed

---

[3] Harkins, W. D., Brown, F. E., and Davies, E. C. H., *J. Am. Chem. Soc.*, **39**, 354–64 (1917).

[4] Langmuir, I., *Met. Chem. Eng.*, **15**, 468–70 (1916).

[5] Langmuir, I., *J. Am. Chem. Soc.*, **39**, 1848–1906 (1917).

between water molecules in the absence of detergent. The final result is that the total force now tending to pull the surface molecules inward is less than previously existed in the case of water alone.

The actual extent to which the surface tension of water is lowered by a given detergent and the rate at which the lowering is effected are dependent on a number of factors, namely, concentration, composition, and temperature. These controlling factors have been studied extensively by many investigators and a large amount of data has been accumulated. Unfortunately, however, much of this data is of little more than qualitative value, particularly for soaps, because

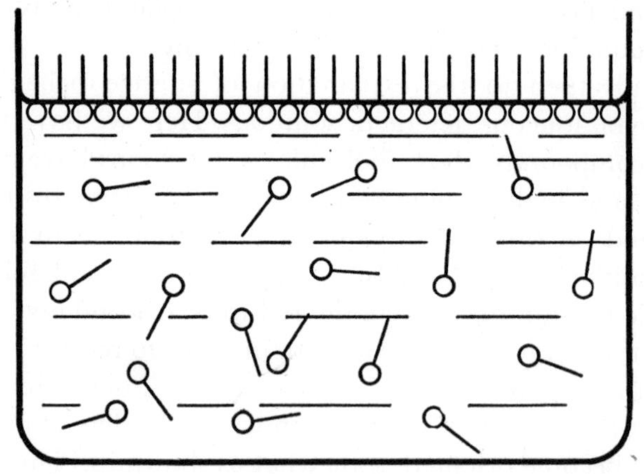

Figure 4–2. Idealized representation of positive adsorption and selective orientation of detergent molecules or fatty ions in a water surface.

of insufficient control of such factors as purity of detergent and exposure to atmospheric carbon dioxide. Therefore, no extensive compilation of surface-tension data will be attempted. The very pronounced influence of even small amounts of impurities will be evident from the discussions to be set forth in Chapter 8.

Any study of surface-tension phenomena must give consideration to two types of surface tension that might be involved, namely, *static* and *dynamic* surface tension. The former represents the tension at equilibrium in an air-liquid interface and the latter the tension at any time subsequent to formation of the interface and prior to attainment of equilibrium. Such differentiation is an acknowledgment that the physico-chemical processes of surface activity involve *time* for their accomplishment. In other words, under a given set of conditions, a definite time is required for attainment of equilibrium in positive adsorption. The effects of the various conditions that control the extent to which surface tension is lowered closely parallel the effects of these conditions on the rate of attainment of static surface tension. Unless otherwise stated, references to surface tension in subsequent discussions will infer static surface tension.

The effects of the composition and concentration of detergent on the extent of lowering of surface tension are illustrated in Figure 4–3

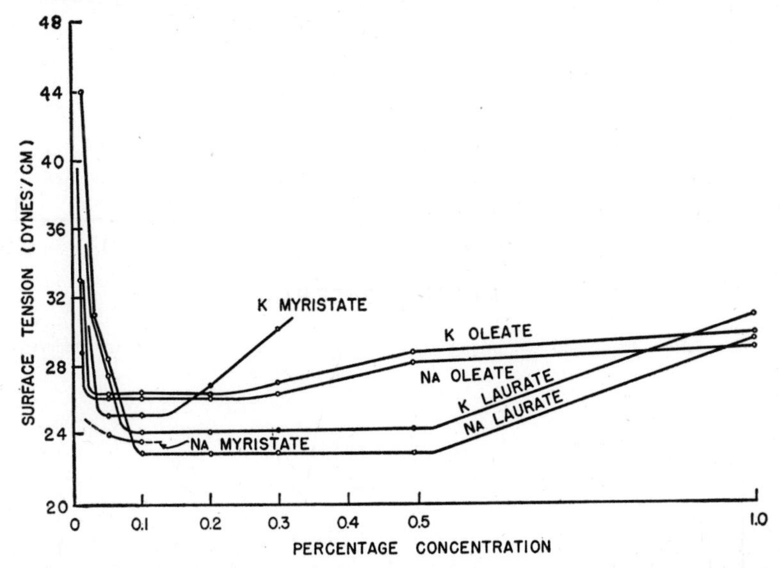

Figure 4–3. Surface tension-concentration curves (20°C) for solutions of several soaps.[6]

from Powney's data[6] for several soaps and in Figures 4–4 to 4–7 from the data of Powney and Addision[7] for several sodium alkyl sulfates. These and data from other sources[8-11] permit the following general summarization:

(a) With increasing concentration of solutions of a given detergent, the surface tension of the solution falls to a minimum, followed by a slight rise at higher concentrations;

[6] Powney, J., *Trans. Faraday Soc.*, **31,** 1510–21 (1935).
[7] Powney, J., and Addison, C. C., *Trans. Faraday Soc.*, **33,** 1243–53 (1937).
[8] Hirose, M., and Shimomura, T., *J. Soc. Chem. Ind. Japan*, **33,** Suppl. binding 337–8 (1930).
[9] Lottermoser, A., and Tesch, W., *Kolloid-Beihefte*, **34,** 339–72 (1931).
[10] Lottermoser, A., and Schladitz, E., *Kolloid-Z.*, **63,** 295–304 (1933).
[11] Lascaray, L., *Kolloid-Z.*, **34,** 73–83 (1924).

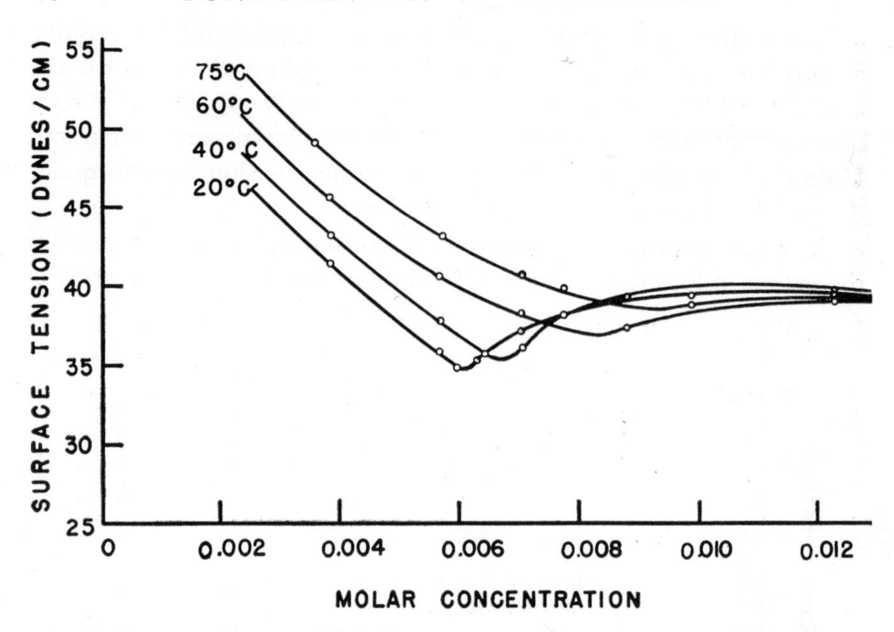

Figure 4–4. Surface tension-concentration curves for sodium dodecyl sulfate at various temperatures.[7]

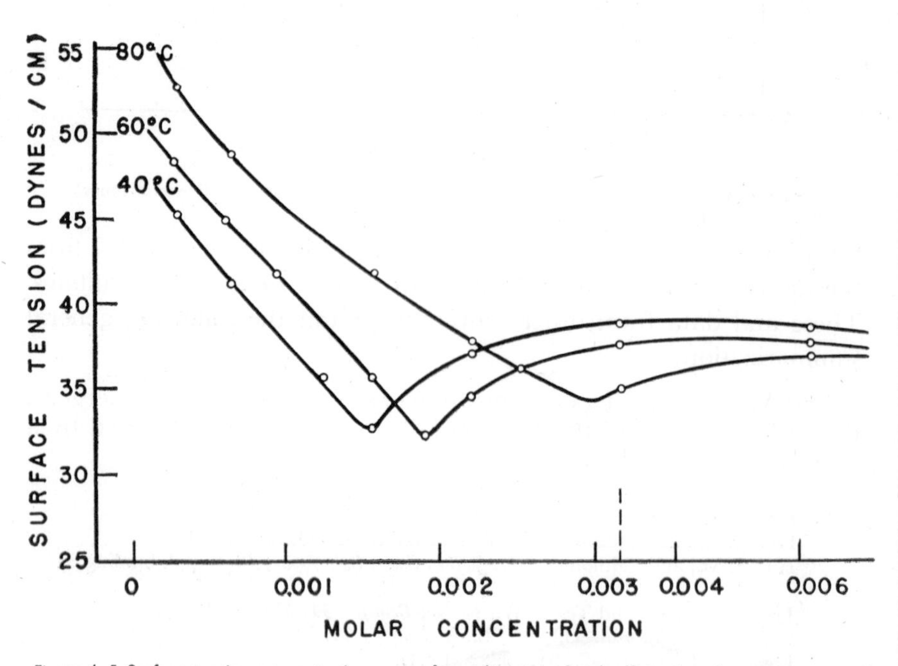

Figure 4–5. Surface tension-concentration curves for sodium tetradecyl sulfate at various temperatures.[7]

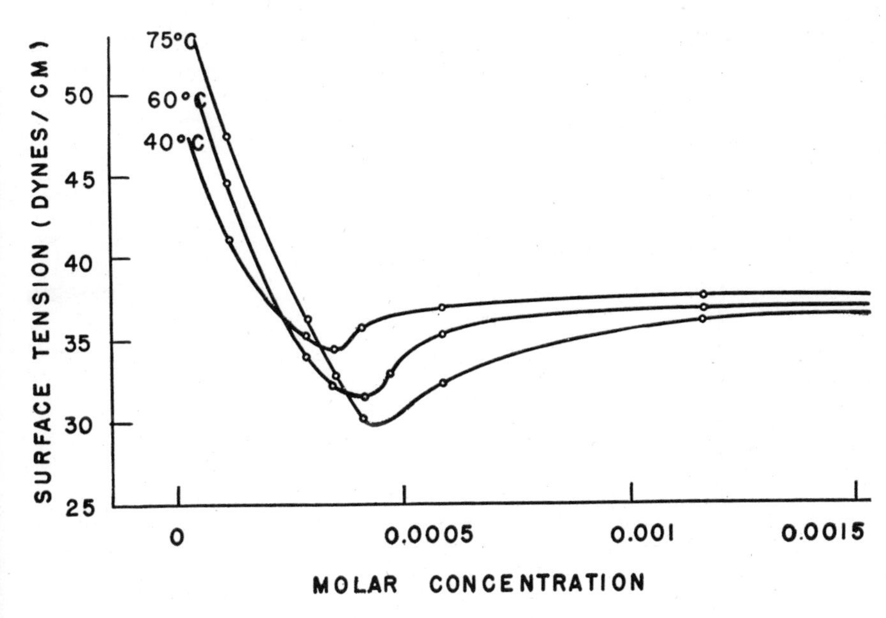

Figure 4–6. Surface tension-concentration curves for sodium hexadecyl sulfate at various temperatures.[7]

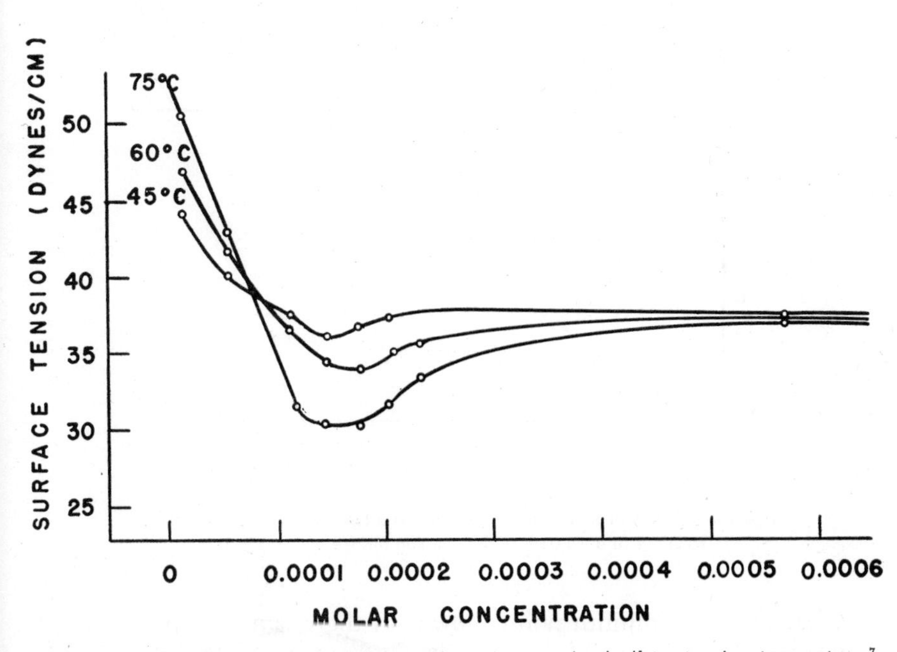

Figure 4–7. Surface tension-concentration curves for sodium octadecyl sulfate at various temperatures.[7]

(b) Between solutions of the detergents of a given homologous series, there is no pronounced difference in the observed surface

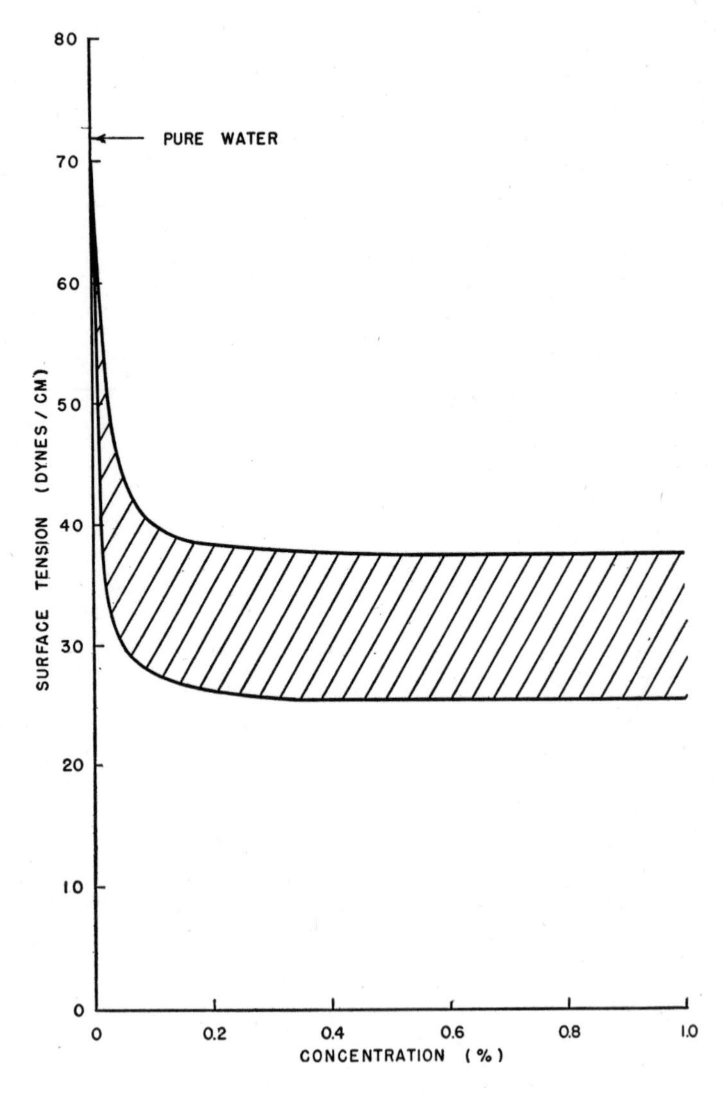

Figure 4–8. Range of surface tension values (as a function of concentration) for solutions of a majority of the common surface-active agents.[12]

tension at the minimum point; however, there is a very pronounced difference in the detergent concentration at which the minimum is attained, this concentration decreasing markedly with increase in molecular weight;

(c) As was the case with hydrolysis, the presence of unsaturation in the hydrocarbon portion of the detergent makes the surface-tension characteristics similar to those of a saturated detergent of lower molecular weight;

(d) As between solutions of sodium and potassium salts with the same anionic group, the surface-tension lowering effected by the latter is somewhat less, but not markedly so;

(e) As between solutions of anionic detergents consisting of the same hydrocarbon group but with different polar groups, the minimum surface tension attainable does not vary appreciably, provided the polar group is attached to the hydrocarbon group in the same position in each case.

In Figure 4–8, Fischer and Gans[12] have well illustrated the fact that there does not exist as wide a variation in surface-tension characteristics with variation in detergent composition as might be expected. This figure, showing the surface-tension values for solutions of a large number of surface-active agents at concentrations up to 1 per cent, illustrates that the tension values fall within a relatively narrow range. This would indicate that, strictly from the standpoint of minimum attainable surface tension, all of the presently common detergents are quite comparable.

Dreger and coworkers[13] have determined the surface tensions of solutions of a homologous and of an isomeric series of purified sodium salts of secondary alcohol sulfates containing from 11 to 19 carbon atoms and for a straight hydrocarbon chain with the sulfate group in various positions, as listed in Table 4–1. The results of these determinations, in comparison to values obtained for primary alcohol sulfates, are shown in Figure 4–9. The surface tensions vary in a regular way toward more pronounced lowering as the length of the hydrocarbon chain is increased for the symmetrical series, as well as when the sulfate group is moved toward the symmetrical position for the series with 15 carbon atoms. These relationships are also indicated in Table 4–2, which shows the concentrations of the alcohol

[12] Fischer, E. K., and Gans, D. M., *Ann. N. Y. Acad. Sci.*, **46**, 371–406 (1946).

[13] Dreger, E. E., Keim, G. I., Miles, G. D., Shedlovsky, L., and Ross, J., *Ind. Eng. Chem.*, **36**, 610–7 (1944).

TABLE 4–1. IDENTIFICATION OF SULFATES IN TABLE 4–2

Compound	*Formula*	*Name, Sodium-*
11–2	$C_9H_{19}CH(OSO_3Na)CH_3$	Undecane-2-sulfate
11–6	$C_5H_{11}CH(OSO_3Na)C_5H_{11}$	Undecane-6-sulfate
13–2	$C_{11}H_{23}CH(OSO_3Na)CH_3$	Tridecane-2-sulfate
13–7	$C_6H_{13}CH(OSO_3Na)C_6H_{13}$	Tridecane-7-sulfate
15–2	$C_{13}H_{27}CH(OSO_3Na)CH_3$	Pentadecane-2-sulfate
15–4	$C_{11}H_{23}CH(OSO_3Na)C_3H_7$	Pentadecane-4-sulfate
15–6	$C_9H_{19}CH(OSO_3Na)C_5H_{11}$	Pentadecane-6-sulfate
15–8	$C_7H_{15}CH(OSO_3Na)C_7H_{15}$	Pentadecane-8-sulfate
17–2	$C_{15}H_{31}CH(OSO_3Na)CH_3$	Heptadecane-2-sulfate
17–9	$C_8H_{17}CH(OSO_3Na)C_8H_{17}$	Heptadecane-9-sulfate
19–2	$C_{17}H_{35}CH(OSO_3Na)CH_3$	Nonadecane-2-sulfate
19–10	$C_9H_{19}CH(OSO_3Na)C_9H_{19}$	Nonadecane-10-sulfate
10–1	$C_9H_{19}CH_2OSO_3Na$	Decyl sulfate (capryl sulfate)
12–1	$C_{11}H_{23}CH_2OSO_3Na$	Dodecyl sulfate (lauryl sulfate)
14–1	$C_{13}H_{27}CH_2OSO_3Na$	Tetradecyl sulfate (myristyl sulfate)
16–1	$C_{15}H_{31}CH_2OSO_3Na$	Hexadecyl sulfate (cetyl sulfate)

sulfates of the different series required to obtain the same lowering of surface tension.

Summary (a) page 45, relative to surface-tension minima, is admittedly controversial and warrants further consideration. There is

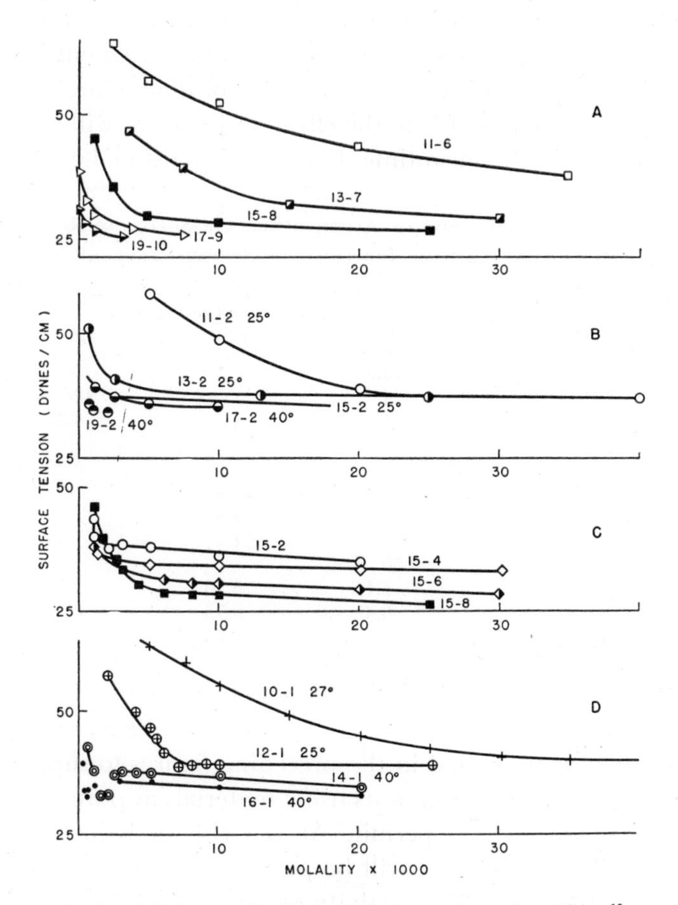

Figure 4–9. Surface tension-concentration curves for sodium sulfates.[13]

A—Sodium *sym-sec*-alcohol sulfates.
B—Sodium *sec*-alcohol sulfates.
C—Isomeric sodium *sec*-pentadecanol sulfates.
D—Sodium *n*-alcohol sulfates.

(See Table 4–2 for description of the compounds.)

much evidence, both in support of this summary and in support of the contention of several investigators, that surface-tension minima should not exist in *pure* detergent-water solutions. As has been pointed out by

Dreger and coworkers,[13,14] Fischer and Gans,[12] Reichenberg,[15] and others, the Gibbs adsorption equation (page 41) does not provide mathematically for minima in the absence of impurities, particularly when the minima are followed by a range of increasing surface tension with increasing concentration. This reversal in action produces the anomaly of an *apparent* range of *negative* adsorption. Discussion of the effects of impurities on the surface activity of detergents, which are potentially quite comparable to the effects of builders, will be reserved until Chapter 8. In the meantime, however, it must be borne in mind

TABLE 4–2. CONCENTRATIONS OF ALCOHOL SULFATE SOLUTIONS REQUIRED TO OBTAIN THE SAME LOWERING OF SURFACE TENSION[13]

Compound[a]	Molal Concentration (× 10³) Required to Obtain Surface Tension of		
	45[b]	40[b]	30[b]
11–6	16.5		
13–7	4		22
15–8	1		4.2
17–9			1
19–10			0.25
11–2		19.4	
13–2		3.6	
15–2		1	
10–1	20		
12–1	5.1	6.50	
14–1	0.4	0.75	
16–1		0.05	

[a] See Table 4–1 for identification of compounds.
[b] In dynes/cm.

that data such as that cited in the preceding figures to support Summary (a) were obtained from tests using materials as pure as presently available methods would permit. At the risk of being accused of avoiding an issue, this point will be pursued no further at this time, on the basis that the surface activity of pure detergents is of interest in the present dissertation only in connection with the later discussions of the surface activity of *built* solutions.

In selecting a detergent to attain a desired surface-tension lowering at a given temperature, particular attention must be paid to the solubility of the detergent at that temperature. To be sure, the surface activity of a detergent is generally higher the greater its molecular

[14] Miles, G. D., *J. Phys. Chem.*, **49**, 71–6 (1945).
[15] Reichenberg, D., *Trans. Faraday Soc.*, **43**, 467–78 (1947).

weight, as pointed out above; however, this statement must also carry the provision that it is true only if the solubility of the detergent at the particular temperature is such as to assure enough dissolved material to attain maximum surface activity. This is well illustrated by such a soap as sodium stearate which, at the same equivalent concentration at higher temperatures, is definitely more surface active than the C_{18} unsaturated soaps and the saturated soaps of shorter chain length. On the other hand, sodium stearate would be useless at a temperature of, say, 20°C because of its low solubility at that temperature. This provision is further illustrated by Figures 4–4 to 4–7, for sodium alkyl sulfates, in which the effects of temperature on surface tension may be seen. Thus, in the case of the alkyl sulfates of lower molecular weight (C_{12} and C_{14}), an increase in temperature does not greatly affect the magnitude of the point of minimum surface tension but decidedly increases the concentration necessary to attain the minimum. On the other hand, in the case of the C_{16} and C_{18} alkyl sulfates, an increase in temperature does not greatly affect the concentration necessary to attain the minimum surface tension but decidedly raises the minimum value.

Hirose and Shimomura,[8] in studying the relationship between the degree of unsaturation of C_{18} soaps and the surface tension of their aqueous solutions, find at high temperatures that the extent of surface tension lowering becomes less the greater the degree of unsaturation, the saturated stearate soap giving the greatest effect. At low temperatures the same tendency is observed with the unsaturated C_{18} soaps, but in this case the stearate soap gives the least lowering of surface tension.

Petrova, Bobuileva, and Nikolaeva[16] find that soaps like tallow soaps, which contain a high proportion of *saturated* fatty acid, have greater surface activity at 60°C than at 20°C, which they attribute to the greater solubility of palmitates and stearates at higher temperatures. Soap solutions containing mostly unsaturated fatty acids, like linseed oil soaps, are more soluble and do not show the temperature effect so strongly. Reed and Tartar[17] have found from a study of the surface tensions of solutions of several sodium alkyl sulfonates that, at 60°C, it is necessary to go to sodium *n*-octadecyl sulfonate (C_{18}) to

[16] Petrova, N. N., Bobuileva, E. N., and Nikolaeva, E. N., "Vsesoyuznuii Nauch.-Issledovatel," *Inst. Zhirov, Untersuchungen über Physikochemie der Waschwirkung*, **1935**, 49–66.

[17] Reed, R. M., and Tartar, H. V., *J. Am. Chem. Soc.*, **58**, 322–32 (1936).

obtain a surface tension comparable to sodium lauryl sulfonate (C_{12}) at 20°C.

Harkins, Davies, and Clark[2] point out that films of solute, such as soap, positively adsorbed at the surface, frequently become saturated and that the rapidity with which this saturation occurs depends on the nature of the solute. Thus, sodium oleate with a C_{18} carbon chain is very highly adsorbed positively in water and the surface tension of its solution rapidly reaches the minimum corresponding to the particular concentration.

It is likely that adsorbed detergent molecules or ions form only unimolecular layers in the surface of a solution and that, as between different members of a homologous series of detergents, the cross-sectional area occupied per adsorbed molecule or ion is constant. In a study of solutions of fatty acids Szyszkowski[18] found that, through a wide range of the over-all concentration c, the capillary elevation y of the solution could be expressed by the equation

$$y = 1 - b \log \frac{c}{a + 1} \tag{2}$$

wherein the constant b depends only on the units employed. Glasstone[19] has developed Szyszkowski's derivation, for *relatively concentrated solutions* of soluble fatty acids, into the form

$$\frac{\gamma}{\gamma_0} = 1 - X \ln \frac{c}{Y} \tag{3}$$

where γ and γ_0 are the surface tensions of solution of concentration c and of pure water, respectively, and X and Y are constants. X is constant for a homologous series of fatty acids of chain lengths from C_2 to C_6; however, Y decreases with increasing chain length. Differentiation of equation (3) with respect to ln c yields

$$\frac{d\gamma}{d \ln c} = -X\gamma_0 \tag{4}$$

and substitution of this value for $d\gamma/d \ln c$ into the approximate Gibbs adsorption equation (equation (1), page 41) gives

$$\Gamma = \frac{X}{RT} \gamma_0 \tag{5}$$

---

[18] Szyszkowski, B. von, *Z. physik. Chem.*, **64**, 385–414 (1908).

[19] Glasstone, S., "Textbook of Physical Chemistry," 2nd ed., pp. 1209–10, New York, D. Van Nostrand Company, Inc., 1946.

Γ represents the excess of surface concentration over bulk concentration, X is the same for any of a series of fatty acids, and γ_0 is constant. Therefore, at the fairly high concentrations to which Glasstone's derivation applies, the excess of fatty acid in the surface likewise becomes constant and independent of the fatty acid. This implies that the surface becomes saturated at sufficiently high bulk concentration and that the saturated surface layer thus formed is only unimolecular in thickness. Glasstone further substantiates that the adsorbed molecules are oriented vertically in the surface and that each molecule occupies a cross-sectional area which is independent of the length of the molecule. By extending Glasstone's derivations for solutions of soluble fatty acids to surface-active agents of higher molecular weight, it may be expected that any molecule or fatty ion of a homologous series will occupy the same cross-sectional area as any other and that, at saturation, the same number of molecules or ions will be present per unit of surface area.

In a study of cationic surface-active agents, Adam and Shute[20] find that the surface tensions of very dilute solutions fall slowly for a week or longer, ultimately reaching very nearly the same tensions as those of more concentrated solutions. Above a critical concentration of 0.001*N* for a 16-C and 0.01*N* for a 12-C hydrocarbon chain, the minimum surface tension value is reached almost at once.

The changes with time of the surface tensions of solutions of sodium cetyl sulfate and sodium lauryl sulfate at different concentrations are illustrated by Figures 4–10 and 4–11, taken from the work of Nutting, Long, and Harkins.[21] It will be noted that for the more dilute solutions of sodium cetyl sulfate the surface tension *vs.* time curves fall rather steeply over a time range of 25 to 50 minutes and then level off almost horizontally. These observations hold generally also for the sodium lauryl sulfate solutions excepting that, below a concentration of about 0.01*N*, the surface tension reaches a broad minimum in time and then slowly rises. In the case of the more concentrated solutions of both agents the initial rate of decrease in surface tension was so fast that the steep portion of the curves was passed before the first measurements were made (at a surface age of 2 to 3 minutes). As was found by Adam and Shute, the range of concentra-

---

[20] Adam, N. K., and Shute, H. L., *Trans. Faraday Soc.*, **34**, 758–65 (1938).
[21] Nutting, G. C., Long, F. A., and Harkins, W. D., *J. Am. Chem. Soc.*, **62**, 1496 501 (1940).

tions that would eventually produce the same ultimate surface-tension lowering was quite broad. Nutting's explanation of the controlling

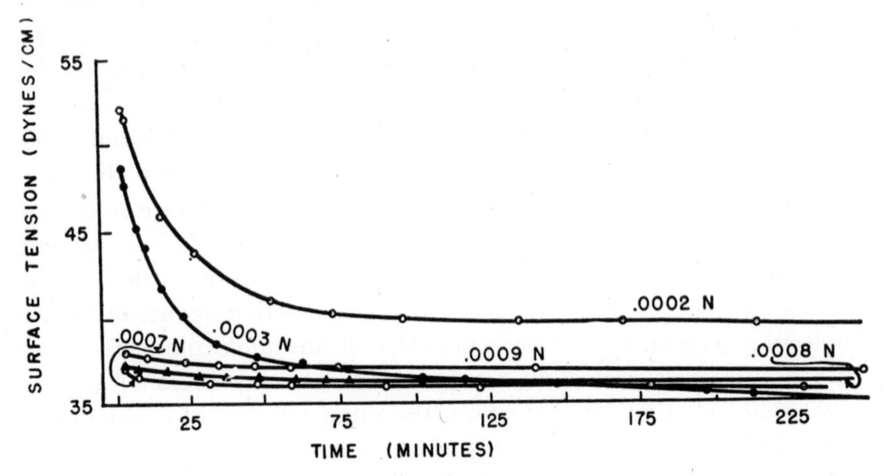

Figure 4–10. Surface tension vs. time curves for various concentrations of sodium cetyl sulfate at 40°C.[21]

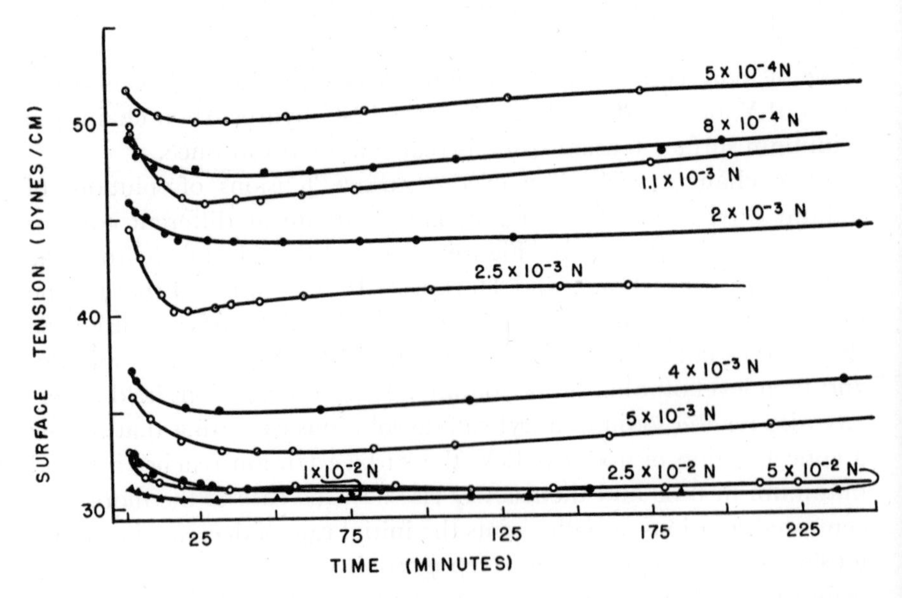

Figure 4–11. Surface tension vs. time curves for various concentrations of sodium lauryl sulfate at 40°C.[21]

factors for rate of attainment of minimum surface tension will be considered in Chapter 8.

The rate of attainment of static surface tension in very dilute solutions or in solutions of the lesser surface-active agents is controlled by two factors. The first is simply that of the time required for the surface-active species to migrate from the bulk of the solution to the air-solution interface until an equilibrium concentration in the interface is reached. The second factor is an electrostatic opposition to positive adsorption. Thus, Doss,[22] in a study of the aging of surfaces of aqueous solutions of cationic agents, has postulated the existence of a potential barrier to account for slow attainment of minimum surface tension in dilute solutions. He considers the first ions to reach the surface to form an electric double layer, the charge on the ions in the surface tending to repel other similarly charged ions attempting to enter the surface.

A little-known factor which may influence the surface tension of aqueous detergent solutions is the action of light. Using sodium oleate solutions of 10^{-4} and 10^{-5} M concentrations, du Noüy[23] reports that solutions prepared and measured in the dark have somewhat higher surface tensions than the same solutions after removal to the light. Further details of this phenomenon are not known to have been studied, nor has a determination been made as to whether this effect is of more than passing interest to detergent processes.

As was pointed out in Chapter 2, commercial soaps are seldom composed of one fatty-acid salt but, rather, are almost always mixtures of several soaps. Our attention must therefore be directed to the surface-tension characteristics of aqueous solutions of mixed detergents, in comparison to the tensions of solutions of single detergents as previously discussed in the present chapter. Perhaps the most extensive studies of such mixtures are those of Mikumo,[24-26] who has shown that, as between two soaps of different molecular weight or different degree of saturation in the same solution, the one of higher molecular weight or greater saturation will be the more positively adsorbed at the surface. He found particularly in soap mixtures where at least one of the soaps was beyond the range of ordinary soaps (such as sodium behenate, C_{22}) that the extent of positive adsorption of the higher soaps was such as to cancel out the surface-active

[22] Doss, K. S. G., *Kolloid-Z.*, **86**, 205–13 (1939).
[23] du Noüy, P. L., *Nature*, **131**, 689 (1933).
[24] Mikumo, J., *J. Soc. Chem. Ind. Japan*, **34**, Suppl. binding, 115–6 (1931).
[25] Mikumo, J., *J. Soc. Chem. Ind. Japan*, **37**, Suppl. binding, 591–3 (1934).
[26] Mikumo, J., *J. Soc. Chem. Ind.*, **52**, 65–8T (1933).

properties of the lower soaps. Lottermoser and Schladitz[10] have shown that the surface tension of solutions of binary soap mixtures lies between those of the pure constituents. Winsor[27] reports that sodium stearate readily displaces sodium *n*-tetradecyl sulfate from the surface layer of aqueous solutions and that the sulfates having the $—OSO_3Na$ group farthest removed from the end of the hydrocarbon chain are most easily displaced. These instances of the effects of one detergent on the surface activity of another immediately indicate a possible analogy between such effects and the effects of the common builders on surface activity.

Interfacial Tension

So far in this discussion of surface activity, we have considered only so-called surface tension or, in other words, the tension of a liquid-vapor interface. Interfacial tension differs from surface tension in that it pertains to the tension of a liquid-liquid or liquid-solid interface, and is thus of more direct significance in detergency than is surface tension alone. The study of interfacial tension at a liquid-solid interface, while it should be of considerable assistance in studies of detergency, has received very little impetus because of lack of suitable means of measurement. Liquid-liquid interfaces, on the other hand, have been studied extensively.

Even though two liquids are completely immiscible (insoluble in each other) when in contact, the surface properties of either one are always affected by the proximity of the other. The interface between two immiscible liquids may be considered to consist of molecules of one liquid "sandwiched" to some extent between molecules of the other. This "sandwiching" affects the inter-molecular attraction of each liquid to such extent that the *interfacial* tension is some component of the surface tensions of the two liquids individually. When two miscible liquids are placed in contact, even though their miscibility is very limited, the equilibrium interfacial tension then becomes some component of the surface tensions of the two liquids when each is saturated with the other.

The tension of a liquid-liquid interface is numerically equal to the work that would be required to increase the area of the interface by one square centimeter. In the case of a liquid A saturated with a

---

[27] Winsor, P. A., *Nature*, **157**, 660 (1946).

liquid B in contact with liquid B saturated with liquid A, a commonly used approximation of the interfacial tension is

$$\gamma_{AB} = \gamma_A - \gamma_B \qquad (6)$$

where γ_A and γ_B are the surface tensions of the respective saturated liquids. Such an approximation has definite and important limitations which are not always recognized. Thus, among other factors, the extent to which the molecules of one liquid "sandwich" themselves between the molecules of the other liquid at the interface has a pronounced effect on the resulting interfacial tension. This may be illustrated by two cases: (a) a highly nonpolar liquid, such as a saturated hydrocarbon oil, and water will form an interface whose tension is closely approximated by equation (6); (b) an organic liquid containing a polar group, such as hydroxyl, carboxyl, carbonyl, or unsaturated carbon-carbon linkage, and substituted organic liquids containing such "water-loving" groups as nitrate, halide, or sulfate, will form with water an interface whose tension is below that predicted by equation (6). The miscibility in case (b) may be very low; the deviation from equation (6) very well may be due to the greater extent of "sandwiching" brought about by the polar groups, with consequent greater penetration of hydrocarbon into the water surface.

Pound[28] has determined the interfacial tensions between a large number of organic liquids and water or various aqueous solutions. He points out that the greater the mutual solubility between the two liquids, the lower is their interfacial tension. It may also be said that the lower the interfacial tension between two liquids, *in the absence of a third component*, the greater is their mutual solubility, from which it follows that the criterion for complete miscibility of two pure liquids in each other is that their interfacial tension be zero or, more properly, that no interface exist between them.

Harkins, Brown, and Davies[3] show that the film of a liquid in contact with water is composed of molecules oriented so that active (or polar) groups are in contact with the water. They show also that the attraction between water and another liquid is one of the important factors in the determination of the solubility of the other liquid in water. In detergency, we are concerned mostly with the characteristics of the interface between water and liquids which, at most, are only

[28] Pound, J. R., *J. Phys. Chem.*, **30**, 791–817 (1926).

slightly miscible with it. In fact, as pointed out later in Chapter 10, many of the organic oily soils encountered in laundering approach complete immiscibility with water.

The interfacial tension between two pure liquids being a function of the degree of miscibility of each in the other, it follows that the interfacial tension between water and slightly miscible oily soil will be high. Water, with a surface tension of 72.8 dynes per cm. at 20°C and, for example, a high paraffinic oil with a surface tension in the range of 30 dynes per cm., would show a relatively high interfacial tension of 40 to 45 dynes per cm. If, however, the surface tension of the water is reduced, as by the addition of a third component such as soap, to, say, approximately the same value as that of the oil, then the interfacial tension of the three-component system will approach zero.

Reduction of the interfacial tension at a water-oil interface by the addition of a third component such as soap cannot be considered to increase the solubility of each liquid in the other. The effect of the addition of the third component, the surface-active agent in this case, is pictured in Figure 4–12, from which it is seen that the soap, in effect, "bonds" the water to the oil to form an interface, by the solubility of the polar end of the soap molecule or fatty anion in the water and of the nonpolar end in the oil.

The illustration given in Figure 4–12 can be extended further to show the effect of the length of the hydrocarbon chain on the interfacial characteristics. If we use as the third component a highly water-soluble fatty-acid salt, such as sodium acetate or sodium propionate, the hydrocarbon end of the molecule or anion is not sufficiently water-repellent to prevent the polar group from pulling the hydrocarbon end entirely into the water phase. In other words, the forces tending to expel the hydrocarbon end are less than the forces tending to disperse it in the solution. The hydrocarbon "film" on the water surface, formed by orientation of the salt molecules or anions, is therefore not pronounced. However, as we ascend the homologous series of fatty-acid salts, the hydrocarbon end of the salt molecule becomes more water-repellent and the hydrocarbon "film" on the water surface becomes more pronounced, that is, the salt becomes more positively adsorbed on the water surface. If, now, an oil is placed on the water, the water presents what might be considered an oil surface and, in effect, an oil-to-oil contact is made with the added oil. Carrying the example farther by supposing that the fatty-acid salt series is ascended on through the soaps, the point is finally reached where the hydrocarbon end of the salt molecule is so water-repellent that the oil layer may "steal" the salt away from the water layer and take it entirely into solution. The interface then again approaches that of a two-component system, with resulting higher interfacial tension.

On the basis that equation (6), page 59, affords a reasonable approximation of the interfacial tension to be anticipated when the surface tensions of two liquids are known, we should expect the inter-

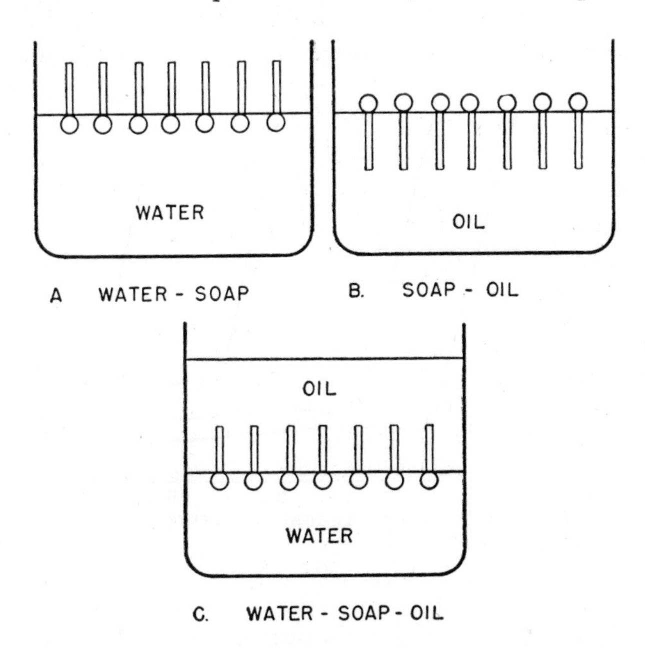

Figure 4–12. Schematic illustration of the action of soap at an oil-water interface.

facial tension between oil and an aqueous solution to more closely approach zero the closer we adjust the surface tension of the solution to that of the oil. This expectation is realized to a considerable extent in actual practice, particularly if the oil is free of reactive material such as fatty acids. By the proper choice of composition, concentration, and temperature, we can "predesign" a detergent solution to approach quite closely a minimum interfacial tension against many oils. However, to effect the reduction of the last few possible dynes of interfacial tension, we must consider some factors other than the true surface tension of each liquid saturated with the other. These factors include the nature of the surface-active species as discussed at the end of this chapter and the phenomenon of electro-capillarity as discussed in Chapter 7. Thus, among other things, the detergent species that most accounts for surface-tension lowering may not always be the same species that effects the last possible lowering of interfacial tension.

The effects of composition and concentration of detergents on the interfacial tension of their aqueous solutions against an oil and also the effect of temperature are quite similar in several respects to the same effects on surface tension and may be summarized as follows:

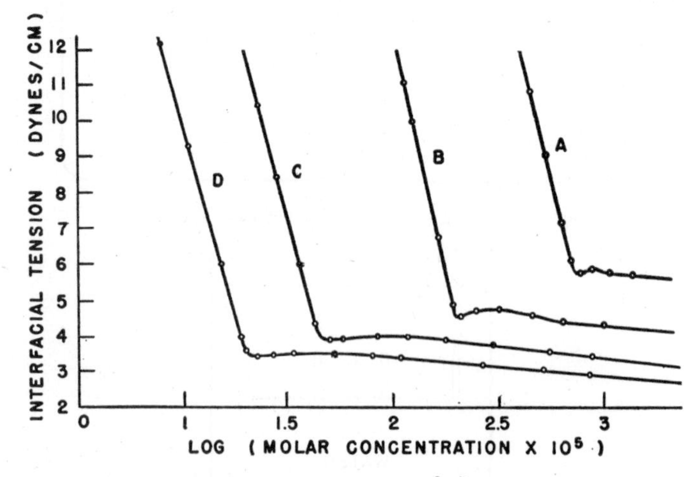

Figure 4–13. Interfacial tensions against xylene at 60°C.[7] Curves A, B, C, and D are for sodium dodecyl, tetradecyl, hexadecyl, and octadecyl sulfates, respectively.

(a) Starting with very dilute solutions and with increasing concentration, the interfacial tension decreases to a minimum; the subsequent increase in interfacial tension with increased detergent concentration is not pronounced, if any (see Figures 4–13 and 4–14 from the data of Powney and Addison[7] for interfacial tension *vs.* concentration of alkyl sodium sulfate solutions against xylene);

(b) In the case of detergents which are quite hydrolyzable, the position of the minimum in the interfacial tension *vs.* concentration curve may be masked by the effect of hydrolysis products (see curves B, C, and D of Figure 8–16, Chapter 8);

(c) Increased molecular weight of the detergent effects a reduction in the concentration necessary to attain the minimum interfacial tension and a reduction in the value of the interfacial tension at the minimum, *provided* hydrolysis is not so extensive as to mask the minimum and bearing in mind that higher temperatures reduce the effect of hydrolysis;

(d) Other factors being constant, an increase in temperature increases the concentration necessary to attain the minimum and also increases the value of the minimum interfacial tension;

(e) As was the case for surface tension, the presence of unsaturation in a soap has the effect of reducing interfacial activity[29]; little difference is to be noted in the interfacial activity of sodium soaps and of potassium soaps of the same fatty acid.

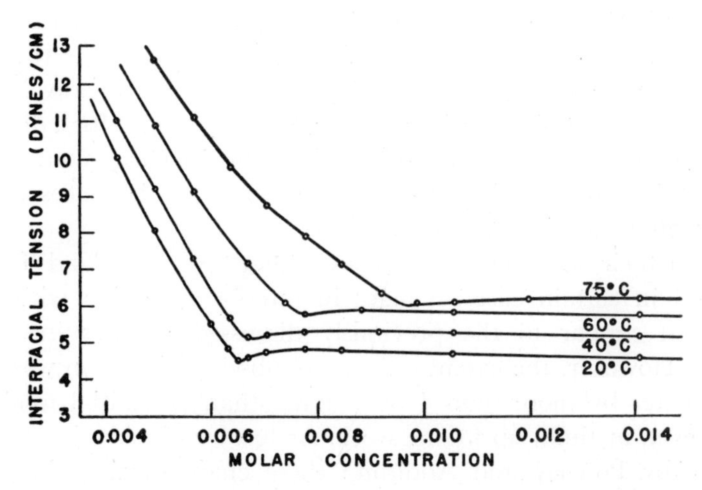

Figure 4–14. Interfacial tension-concentration curves (against xylene) for sodium dodecyl sulfate at various temperatures.[7]

Another very significant factor in interfacial tension is the nature of the oil phase, even though the oil may be chemically inert. Thus, the ability of the oil to dissolve surface-active material from the interface, as determined by such factors as molecular weight and chemical constitution, has a definite bearing both on the magnitude of the interfacial tension and on the time required to attain equilibrium.

Davis and Bartell,[30] in studying the interfacial tension between

---

[29] Powney, J., and Addison, C. C., *Trans. Faraday Soc.*, **34**, 356–63 (1938).
[30] Davis, J. K., and Bartell, F. E., *J. Phys. Chem.*, **47**, 40–50 (1943).

heptane and aqueous sodium laurate solutions, found that there is a pronounced lowering of the interfacial tension with time, due to hydrolysis of the sodium laurate and diffusion of lauric acid into the heptane. This, however, is not the only or necessarily the most significant way in which time influences interfacial tension. As has been previously pointed out, the maximum extent of lowering of the surface tension of water by dissolved detergent is not reached until the detergent has had time to become adsorbed to the fullest extent in the water surface. Further, the ultimate interfacial tension developed between two liquids is that between the liquids when each is saturated with the other. Therefore, the time required to attain this mutual saturation governs the time of attainment of the final interfacial tension. Powney and Addison[29] consider that the first of these, that is, the time to attain saturation of the interface, is relatively short, while the time required for oil-soluble species to travel across the interface to attain equilibrium may be quite long. Much difficulty is encountered in attempts to correlate interfacial-tension data of different investigators because of failure in many instances to take into account these time factors.

From their study of sodium alkyl sulfonates, Reed and Tartar[17] observed in certain cases changes in interfacial tension with time, somewhat similar to the previously discussed changes in surface tension. However, the extent of increase observed by them was in no case greater by more than 4 or 5 dynes than the initial minimum value, even at times up to one week. As has been pointed out subsequently by Powney and Addison,[7] these changes might be partly attributable to foreign material dissolved from the container.

Fischer and Gans[12] have summarized the interfacial-tension values between mineral oil and water for the majority of surface-active agents as shown in Figure 4–15.

Liquid-Solid Interfaces

As previously mentioned, direct means for the measurement of liquid-solid interfacial tensions have not been successfully developed. This difficulty was first illustrated by Young[31] in 1805, who derived an expression for the equilibrium conditions obtaining in a solid-liquid system in terms of the contact angle between the solid and

[31] Young, T., *Phil. Trans.*, **65** (1805).

liquid, the surface tensions of the solid and of the liquid, and the solid-liquid interfacial tension, as

$$\gamma_s - \gamma_{sl} = \gamma_l \cos \theta \tag{7}$$

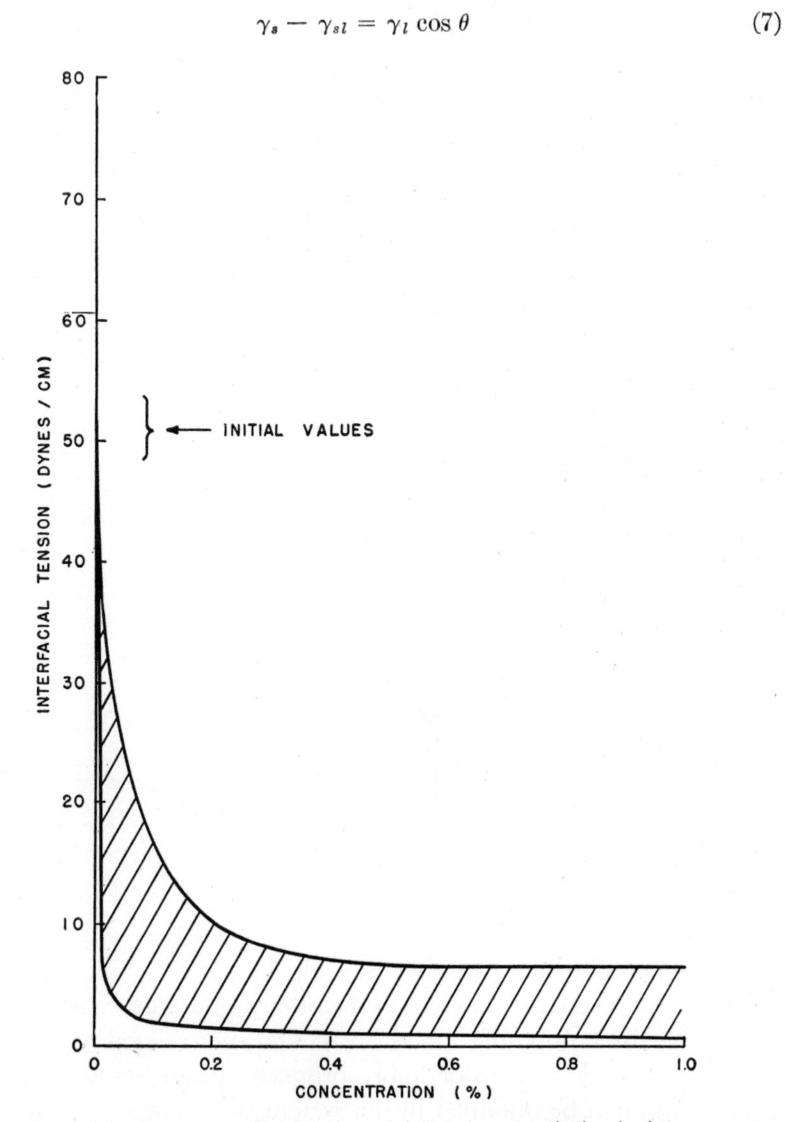

Figure 4–15. Range of interfacial tension values between mineral oil and solutions of a majority of the common surface-active agents.[12]

wherein γ represents surface or interfacial tension, subscripts s and l denote solid and liquid, respectively, and θ is the angle of contact.

This relationship is illustrated in Figure 4–16. At equilibrium, the forces, γ_s, γ_l, and γ_{sl}, must balance, and the relationship between γ_s and the sum of γ_l and γ_{sl} governs the size of angle θ. Thus, if the surface tension of the liquid is sufficiently high with reference to that of the solid, angle θ will be large. Arbitrarily, it may be said that: If θ is greater than 90°, the liquid does not wet the solid; if θ is less than 90° but greater than 0°, the liquid partially wets the solid; and if there is no contact angle, the liquid completely wets the solid. A hypothetical case in addition to these has been pointed out recently by Fischer and Gans[12] because of the erroneous use of the term "zero contact angle." They show that, in the case of complete wetting and

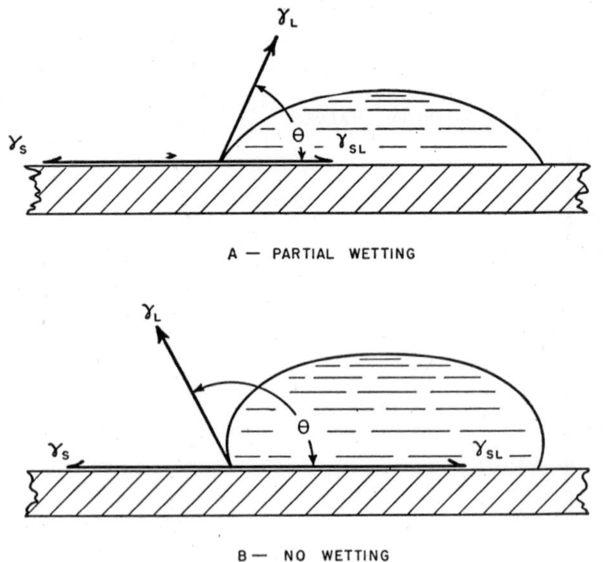

Figure 4–16. Illustration of the equilibrium conditions at a liquid-solid interface in terms of contact angle.

spreading, equilibrium cannot exist and, thus, under such a condition there can be no contact angle. On the other hand, in the rare instance where the three opposing forces are *just* so interrelated as to reduce the contact angle to zero or, more properly, to an infinitesimal size, equilibrium can be obtained in the system but complete wetting still will not occur.

Antonoff[32] has developed what he considers to be a quantitative method for indirectly measuring the surface tension of solids. His

[32] Antonoff, G., *J. Phys. & Colloid Chem.*, **52**, 969–75 (1948).

method is based on the use of liquids or pastes, the surface tension of which is approximately that of the solid in question. If its surface tension is higher than that of the solid, the liquid will not wet the solid; if the surface tension is lower, the liquid adheres to the solid. By changing the surface tension of the liquid in small increments through addition of a third material (usually a finely powdered pigment), he reports reaching a point where the liquid just begins to wet the solid. At this point Antonoff considers the surface tension of the solid to equal that of the liquid, and the latter value can be determined by known methods. Antonoff presents considerable theoretical discussion to substantiate his method in the face of adverse criticism by several other workers. One of the most vulnerable points in his technique is his method of measuring the surface tension of the special pastes for solids of high surface tension.

Antonoff reports a surface tension of 130 dynes per cm. for ordinary glass. He considers that the expression of the form,

$$\gamma_A - \gamma_B = \gamma_{AB}$$

for liquid-liquid systems, is valid at a solid-liquid interface, with the provision that there is no equilibrium in the latter case and the symbols refer to tensions of pure substances at the moment of contact. For γ_{AB} in a solid-liquid system to be zero (Antonoff's condition of complete wetting), γ_A (for, say, the solid) and γ_B must be equal. In the case of pure water we have imperfect wetting of glass but this condition is readily overcome by the use of a suitable surface-active agent. From our previous discussions of surface tension we know that the minimum surface tension of, say, a soap solution is about 30 dynes per cm. Substituting this and Antonoff's value for the surface tension of glass into the above equation we would seem to have

$$130 \text{ dynes/cm. (for glass)} - 30 \text{ dynes/cm. (for soap solution)} = 100 \text{ dynes/cm.}$$

for the interfacial tension. We know, however, that because of the known existence of perfect wetting such cannot be the case. Here then we have apparent evidence of not only reduction of the surface tension of water by soap but also of the simultaneous and considerably greater reduction of the surface tension of the glass by soap. Probably what we are observing in such a case is not a solution-glass interface but, rather, a solution-soap interface, with the soap being adsorbed on the glass surface. Antonoff's method of deriving the surface tension of solids may be subject to a similar limitation, namely, that he is not dealing with a true paste-*solid* interface.

We turn again to the limitation of Young's expression, namely, the difficulty of measuring the surface tension of a solid. This limitation has been circumvented in another way by the use of the free surface-energy relationships at solid-liquid interfaces. Thus, Dupré in 1869 showed that when a solid-liquid interface is decreased by 1 sq. cm. the free surfaces of both the solid and the liquid are increased by 1 sq. cm. Now, an amount of work numerically equal to the surface tension of a liquid or a solid is required to increase the free surface area of either by 1 sq. cm. The net work done in the process is then expressed by

$$W_{sl} = \gamma_s + \gamma_l - \gamma_{sl} \tag{8}$$

This is the work of overcoming the adhesion between 1 sq. cm. of liquid and 1 sq. cm. of solid and is called the *work of adhesion.*

Considering only the surface tension of the solid and the solid-liquid interfacial tension, Freundlich[33] has derived a new term, *adhesion tension*, *A*, which he defines as

$$A = \gamma_s - \gamma_{sl} \tag{9}$$

The greater the attraction between a solid and a liquid, the smaller will be the interfacial tension, γ_{sl}, when these phases are brought in contact, and the greater will be the adhesion tension, *A*, between the two phases. Thus, adhesion tension is a direct measure of the attraction of a solid for a liquid and, likewise, of its wettability by the liquid. Bartell and Osterhof[34, 35] consider Freundlich's adhesion tension as being numerically equal to the work of *immersion*—that is, the work involved in the substitution of a solid-liquid interface for a solid-air interface. In other words, they consider the free surface energy of a solid in air, γ_s, minus the free surface energy of a solid in liquid, γ_{sl}, to be a measure of the energy expended in the process of wetting. However, it must be remembered that Bartell's and Osterhof's interpretation of Freundlich's adhesion tension applies only when the

[33] Freundlich, H., "Colloid and Capillary Chemistry," translated by H. S. Hatfield, New York, E. P. Dutton and Company, 1926.

[34] Bartell, F. E., and Osterhof, H. J., *Ind. Eng. Chem.*, **19**, 1277–80 (1927).

[35] Bartell, F. E., *Ind. Eng. Chem.*, **33**, 737–40 (1941).

solid is *immersed* in the liquid. In the case represented by Figure 4–16, the work of wetting also must take into consideration the free surface energy of the liquid against air, as in Dupré's equation.

Bartell and Osterhof express the adhesion tension in the form

$$A_{sl} = \gamma_s - \gamma_{sl} = \gamma_l \cos\theta \text{ (when } \theta > 0) \quad (10)$$

When there is no angle of contact, as in complete wetting, this equation is indeterminable, and the more general expression,

$$\gamma_s - \gamma_{sl} = K\gamma_l \quad (11)$$

must be resorted to. Since the magnitude of the contact angle is determined by the force of adhesion between solid and liquid, K may be termed the "adhesion constant." They present a test method whereby K can be measured even in those cases where there is no angle of contact, from which it is possible to calculate the adhesion tension.

Reinders,[36] in 1913, reported a study of liquid-solid interfaces using powders and the distribution thereof between two mutually insoluble liquids such as water and oil. He has pointed out that the distribution of the solid will depend upon three interfacial tensions, that between the solid and the water (γ_{sw}), that between the water and the oil (γ_{wo}), and that between the solid and the oil (γ_{so}). He proposes a theorem, as follows:

(a) If (γ_{so}) > (γ_{wo}) + (γ_{sw}), the solid will remain suspended in the water;

(b) If (γ_{sw}) > (γ_{wo}) + (γ_{so}), the solid will leave the water and go into the oil phase;

(c) If (γ_{wo}) > (γ_{sw}) + (γ_{so}), or if none of the three interfacial tensions is greater than the sum of the other two the solid particles will collect at the boundary between the water and the oil.

Gortner[37] points out that the conditions of Reinders' theorem will also hold for a particle which comes in contact with a film of oil instead of a layer or a globule of oil, and will determine whether or not the particle is wetted by the oil film. It will be wetted if (γ_{sw}) is greater than the other two interfacial tensions. It will remain in contact with the oil film if (γ_{wo}) is greater than the other two interfacial tensions.

---

[36] Reinders, W. von, *Kolloid-Z.*, **13**, 235–41 (1913).

[37] Gortner, R. A., "Outlines of Biochemistry," p. 155, New York, John Wiley & Sons, Inc., 1929.

It will not be wetted if (γ_{so}) is greater than the other two interfacial tensions. Although Reinders' theorem is fundamentally sound, it furnishes only qualitative data because of experimental difficulties encountered in working with powdered solids.

The application of these energy relationships at liquid-solid interfaces to the phenomenon of wetting and spreading, and the role played by various detergents therein, will be considered in the next chapter.

Identity of the Surface-active Species

The various forms in which soaps and synthetic detergents may exist in solution have been discussed, as well as the various manifestations of the surface-active nature of these materials. However, up to this point specific discussion of the particular form or species which accounts for the maximum surface activity of detergent solutions has been intentionally avoided. There is considerable disagreement or, perhaps, lack of accurate understanding as to the identity of this species. Be that as it may, the true identity of the species and a knowledge of how we may best assure its presence in the optimum amount are of the utmost importance to the success of any detergent process.

Having determined that solutions of soaps and of ionic synthetic detergents may contain simple ions, simple molecules, and micelles, in proportions depending on the prevailing conditions of composition, concentration, and temperature, the problem now becomes one of determining which of these species is the most surface-active, that is, which is most positively adsorbed at a surface or interface. For at least part of the substantiation of our discussion of this subject it will be necessary that we anticipate some of the discussion that follows in Chapter 8 on the influence of builders on the composition and surface activity of detergent solutions.

In general, there are two "schools of thought" on the nature of the surface-active species. The first favors micelles; the second favors either the detergent molecule or one or another of the products of its hydrolysis or dissociation.

In considering the manner in which soaps lower the surface tension of water, Walker[38] considers it necessary to assume that the lowering is not solely proportional to the concentration of the surface layer, but also depends on the *size* of the colloid particles. He postu-

[38] Walker, E. E., *J. Chem. Soc.*, **119**, 1521–37 (1921).

lates two types of micelles, one very small and very surface-active and the other quite large and only slightly surface-active. In very dilute solutions he ascribes the small decrease in surface tension with increased concentration to the preponderance of simple ions and simple molecules in the solution. On increasing the concentration further, the equilibrium within the system is shifted toward the formation of more of the small surface-active micelles, with consequent pronounced reduction in surface tension. Further increase in concentration finally results in the formation of a preponderance of the larger, less surface-active micelles, with corresponding increase in surface tension.

Lottermoser and Stoll[39] have attempted a correlation of Walker's postulation with McBain's work on the composition of soap solutions. Assuming that McBain's "neutral" micelles correspond to Walker's small, surface-active micelles and that McBain's ionic micelles correspond to Walker's larger, less surface-active micelles, Lottermoser and Stoll arrived at an explanation of the surface-tension behavior they found in a study of various metal alkyl sulfate solutions. From a study of the surface activity of alkyl sulfate salts, Reed and Tartar[17] favor this postulation. They consider that the changes of surface tension with time noted in their work and the large positive temperature coefficient of surface tension can best be explained by assuming that the surface-active unit in the solution is some aggregated particle and not a simple ion.

At this point attention should be directed to an apparent inconsistency in the reasoning of Lottermoser and Stoll. Bearing in mind a characteristic surface tension-concentration curve for, say, soap solution, we have with increasing concentration progressively greater surface activity until a minimum surface tension is reached, after which we have a range of a constant or progressively lessening surface activity. According to McBain, Ekwall, and others, micelle formation through such a transition in concentration follows the order—small ionic micelle, larger ionic micelle, still larger and progressively more neutral micelle. This is just the opposite of the transition pictured by Lottermoser and Stoll in their correlation.

In 1925, Harkins and Clark[40] considered that the surface-active species in the surface film of unbuilt sodium nonylate solution are

---

[39] Lottermoser, A., and Stoll, F., *Kolloid-Z.*, **63**, 49–61 (1933).
[40] Harkins, W. D., and Clark, G. L., *J. Am. Chem. Soc.*, **47**, 1854–6 (1925).

sodium nonylate molecules together with nonylic acid produced by hydrolysis, with the latter being the more active of the two. By way of explaining the increase in surface tension of such solutions by the addition of small amounts of sodium hydroxide, they considered the repression of hydrolysis to decrease the concentration of nonylic acid in the surface film.

Murray[41] assumes that three kinds of particles exist in detergent solutions, namely, ionic micelles, simple metallic cations, and simple long-chain anions. Starting with a very dilute solution, the number of simple anions present in solution will increase with increasing over-all concentration of detergent to a maximum point, after which there will be a decrease due to their rapidly increasing aggregation to form ionic micelles. He considers that the surface-tension minimum occurs at the point where the concentration of anions is greatest prior to their aggregation to form micelles and concludes therefore that the simple long-chain anions are the surface-active units of consequence in such solutions. Powney and Addison[7] consider that confirmation of Murray's postulation is obtained from the close resemblance between the form of the surface and interfacial-tension curves and that of the theoretical curve for concentration of long-chain anions versus total concentration of detergent. Howell and Robinson[42] further point out that, up to the critical concentration, the alkyl sulfate detergents are completely dissociated (as is to be expected with such relatively strong electrolytes) and that it is in this concentration range where surface activity attains its maximum value. Powney and Addison consider that the slight variations in tension at concentrations above the critical can be regarded as due to micelles only in so far as they affect the concentration of the simple long-chain anions present.

Adam,[43] in reviewing the surface tension of soap solutions, proposes that free fatty acid or acid soap should be more strongly adsorbed in the surface film than neutral soap because of the fact that a —COOH group has much less affinity for water than a —COONa group. This is further demonstrated by actual analysis of foam from

---

[41] Murray, R. C., *Trans. Faraday Soc.*, **31**, 207 (1935).

[42] Howell, O. R., and Robinson, H. G. B., *Proc. Roy. Soc.* (London), **A155**, 386–406 (1936).

[43] Adam, N. K., *Trans. Faraday Soc.*, **32**, 653–6 (1936).

soap solutions by Laing and coworkers.[44, 45] Adam considers that the pronounced lowering of surface tension by very low concentrations of soap is indicative that fatty acid or acid soap is exceedingly strongly adsorbed and that the influence of hydrolysis is thus greatly magnified in the surface layer.

Powney and Addison[29] conclude that the species in soap solutions mainly responsible for the *interfacial* activity at an oil-solution interface is different in character from that responsible for the activity at an air-solution interface. Their reason for this differentiation is the fact that the addition of a small amount of alkali may *raise* the surface tension of a soap solution whereas the addition simultaneously *lowers* the interfacial tension against an oil. On the basis that fatty acid or acid soap is the active species in the case of surface tension, they point out that, in an oil-solution system, "the appreciable solubility of the acid soap in the oil phase would virtually prevent the formation of any organized boundary film of this component." They consider it probable therefore that the active species for interfacial-tension lowering, in solutions of both soaps and anionic synthetics, is either long-chain anions or molecules. In substantiation of this they point to the fact that alkyl sulfates exhibit minima in interfacial tension at lower concentrations than do the corresponding soaps, a fact they attribute to all the alkyl sulfate being present in the surface-active state at the concentrations studied.

Mees[46] suggests that the products of dissociation formed in aqueous soap solutions are not essential for the lowering of the interfacial tension between water and oil. He considers it to be more likely that the soap molecule as such is essential in this regard according to its heteropolar character and its resulting orientation. However, his argument does not preclude the products of dissociation and hydrolysis from fulfilling a similar function, as they are likewise highly heteropolar and oriented. Alexander[47] points out that the soap micelles may be considered as reservoirs supplying surface-active soap molecules to the surface or interface as needed.

---

[44] Laing, M. E., *Proc. Roy. Soc.* (London), **109A**, 28–34 (1925).

[45] Laing, M. E., McBain, J. W., and Harrison, E. W., "Colloid Symposium Monograph," Vol. **6**, pp. 63–72, New York, Chemical Catalog Company, Inc., (Reinhold Publishing Corp.), 1928.

[46] Mees, R. T. A., *Chem. Weekblad*, **20**, 302–4 (1923).

[47] Alexander, A. E., *Trans. Faraday Soc.*, **38**, 54–63 (1942).

Each of the two foregoing contradictory general viewpoints appears to have merit. In retrospect the points set forth below should be considered.

(a) The presence of ions, soap molecules, fatty acid, "acid soap," and micelles in water solutions of soap under certain conditions is well established. The differentiation between neutral micelles and ionic micelles is not conclusive and is probably a matter of *degree* of charge on the micelle with the micelle becoming more neutral as the over-all concentration increases. In other words, it is rather difficult to conceive, at laundry concentrations of the order of 0.2 per cent, that the conditions which define the existence of ionic micelles would permit the simultaneous existence of entirely neutral micelles. Quite to the contrary, thermodynamic considerations in a colloidal system favor uniformity of the particles with respect to both size and composition.

(b) Due to the fact that inorganic salts and their ions are only weakly surface-active at best, as will be shown in Chapter 6, consideration of the cations from soap and from other anionic detergents as being surface-active entities can be eliminated from the present discussion.

(c) It is not reasonable to expect that fatty anions, molecules of fatty acid and neutral and "acid" soap, and micelles will act to a like degree as surface-active species, because of differences in size, chemical composition, and heteropolarity. On the other hand, it can be reasonably expected that all five species will be positively adsorbed, although to different extents, at the solution surface or at an interface.

(d) The relative extent to which fatty anions, molecules of fatty acid, neutral soap, "acid soap," or micelles are positively adsorbed at a given time depends, first, on the relative degrees to which they can reduce the surface or interfacial free energy; second, on their relative concentrations in the bulk of the solution; and finally, on their relative rates of diffusion from the bulk of the solution to the surface or interface. Consideration of these factors requires a review of the physical condition of the five species. Briefly, we have a long-chain fatty ion with a hydrophilic $—COO^-$ or similar group; a long-chain neutral molecule with a hydrophilic —COONa or similar group; a long-chain fatty acid molecule with a hydrophilic —COOH group; an acid soap "molecule"; and a micelle. The latter two species may in several respects be considered as varying combinations of the first three. Thus, the micelle is an aggregation of varying proportions of fatty ions and neutral molecules and the acid soap "molecule" may very well be a solubilization product or sorption compound of fatty acid with micelles.

As between a fatty acid molecule, a fatty ion, and a neutral soap molecule, we should expect the relative inclination to positive adsorption to descend in that order. This is predicated on the ascending order of "hydrophilicity" of the active group in each species being: (1) —COOH; (2) —COO^-; (3) —COONa. However, this order cannot necessarily be transposed to any actual soap solution, wherein the concentrations of the three species may differ widely. The highest *relative* concentration of fatty acid occurs in very dilute soap solutions but, as the over-all concentration increases, the fatty-acid concentration becomes less significant in comparison to that of the fatty ion or, at higher concentrations, even to that of undissociated soap molecules. Likewise, with increasing over-all concentration, the relative concentration of fatty ion becomes less. In very dilute solutions, a much greater tendency for positive adsorption of fatty acid molecules might cause more fatty acid to be adsorbed in a surface than fatty ion, even though the latter might be present in the solution in greater number. However, it is not known how far this condition might hold as the over-all concentration is increased. At the over-all concentration where maximum surface activity is realized and where the *relative* concentration of fatty acid must be very low, it is questionable whether fatty acid is present in the surface in sufficient amount to be a significant species. The species most likely to be present in the surface in the greater amount at the over-all concentration which gives maximum surface activity is the fatty anion, based on the likelihood that the fatty anion is available to the greatest extent in the bulk of the solution. On the other hand, there is the unconfirmed possibility that undissociated soap molecules also may be significant at such concentrations.

Going next to the somewhat "special" case of a solution-oil interface, we should expect the conditions which initially cause the various species to be adsorbed on the solution side of the interface to be the same as those which cause them to be adsorbed at a solution-air interface. However, once a particular species has entered the solution-oil interface, the conditions prevailing upon it are different from those prevailing at a solution-air interface. Particularly, there is a tendency for the oil phase to "steal" the detergent from the interface. We should expect the descending order of ease of crossing the interface, to pass into solution in the oil phase, to be: fatty acid; fatty ion; undissociated soap molecule. Perhaps, as postulated by Powney and Addison,[29] this ease of migration across the interface plays some part in the final interfacial-tension value; however, our present knowledge does not permit a conclusion as to the significance of such a part.

Next we must consider the micelle and its place in this picture. Composed as it is of fatty anions and neutral soap molecules aggregated together, in effect, for the very reason of partially reducing the free surface energy of its components, the micelle should be considerably less surface-active than the summation of the same properties for all the individual anions and molecules of which it is composed. If we consider the simple anions and molecules as individual units and the micelle as an aggregate of such units, then we may say that, *per unit*, the micelle is the least surface-active of the three species. This same argument probably holds also for "acid soap."

Referring briefly to the case of anionic synthetic detergents, which is special because of the absence of hydrolysis, the surface-active species are limited to three, namely, fatty anion, molecule, and micelle. Applying the reasoning set forth above for soap, where applicable, the fatty ion should be more surface-active than the molecule or micelle. However, we are again faced with the probability that the relative concentrations of each in the bulk of the solution also affect the relative numbers of each species adsorbed at a surface or interface. By anticipating later discussions, we know that the effects of electrolyte builders on both the surface tension and the interfacial tension of solutions of anionic synthetic detergents are quite comparable, namely, a pronounced lowering in each case. Any postulation which suggests one species for *surface* activity and another species for *interfacial* activity must take this into account. There is no reason for builder favoring the formation of fatty ions in the first instance and the formation of undissociated detergent molecules in the second instance, as would be necessary if these two species were to account for greater *surface* activity and *interfacial* activity, respectively.

Past discussions of surface activity-concentration data have used the total amount of added surface-active agent as the criterion for comparison. Perhaps such discussions would have been strengthened and the conclusions drawn therefrom modified if concentration could have been measured and expressed in terms of that of each of the surface-active species known to be present in the system. Unfortunately, at least for our present discussion, the surface- and interfacial-tension minima for solutions of most soaps or, in other words, the points of maximum surface activity, are "masked" by hydrolysis products. The presence of products of hydrolysis in considerable amounts greatly complicates the problem of identifying and estimating the amount of the various species in a soap solution. Consequently, much of the needed study of soap solutions has been forsaken in recent years for the less difficult studies of solutions of unhydrolyzed synthetic detergents. Conclusive evidence as to the relative surface activities of the various species and as to the conditions which determine the extents to which the various species exert their surface activity can only be obtained by further study of the problem.

5

Actions of Unbuilt Detergent Solutions

Based upon their nature and properties as discussed in Chapter 4, detergent solutions are capable of several typical actions against or with another phase, as follows: *wetting and spreading* on liquid or solid surfaces, such as on liquid or solid soil or on fabrics; *foaming* with gases; *emulsification* of oily material; *peptization* of aggregates of solid particles; *solubilization* of water-insoluble liquids and solids; *ion exchange*; and *stabilization* of dispersed systems. These actions will be discussed in order in the present chapter.

Wetting and Spreading

Cooper and Nuttall[1] in 1915 introduced the concept of *spreading coefficient* in determining the wetting powers of agricultural sprays and dips, and their findings are probably applicable to a considerable extent to laundry conditions. They define the spreading coefficient S of liquid A over liquid B by the equation

$$S = \gamma_B - (\gamma_A + \gamma_{AB}) \tag{1}$$

A positive value for S indicates spontaneous spreading to a thin film. They studied various soap solutions against a "liquid vaseline" surface and found that soaps do not differ markedly from one another at all concentrations tested (up to 2 per cent fatty acid content).

Harkins and Feldman[2] present the following as the criterion for the spreading of a liquid on another liquid or on a solid:

---

[1] Cooper, W. F., and Nuttall, W. H., *J. Agr. Sci.*, **7**, 219–39 (1915).
[2] Harkins, W. D., and Feldman, A., *J. Am. Chem. Soc.*, **44**, 2665–85 (1922).

A liquid will spread if its work of surface cohesion, W_c, is less, and will not spread if its work of surface cohesion is greater, than its work of adhesion, W_a, with respect to the surface of the liquid or solid upon which the spreading is to occur.

The free energy decrease S which occurs in spreading is given by the expression

$$S = \gamma_B - (\gamma_A + \gamma_{AB}) \tag{2}$$

wherein B denotes the liquid on which spreading is to take place and A is the liquid which is to spread. It will be seen that this equation is similar to equation (1) and that Cooper's and Nuttall's spreading coefficient is synonymous with the free energy decrease in spreading. Using Dupré's equation for the work of adhesion,

$$W_a = \gamma_A + \gamma_B - \gamma_{AB} \tag{3}$$

and the expression

$$W_c = 2\gamma_A \tag{4}$$

for the work of cohesion that is necessary to create inside of the spreading liquid an area equal to 2 sq. cm., they derive, by combination of the foregoing three equations, the general expression for spreading coefficient, as

$$S = W_a - W_c \tag{5}$$

This expression states, in effect, that spreading occurs if the adhesion between the two liquids is greater than the cohesion in the liquid which is in the position for spreading. They point out also that, because a liquid A spreads upon a liquid B, it does not necessarily follow that liquid B will spread upon liquid A. The foregoing expressions and generalizations hold also for liquid-on-solid systems, if terms for the solid are substituted for terms for liquid B.

Harkins and Feldman show that, since the free surface energy of almost all inorganic solids is high, W_c for these solids is high and W_a is also high with reference to practically all liquid substances. They consider then that, since the work of cohesion in water and organic liquids is, in general, low, the values of the coefficient of spreading of these liquids upon such solids should be positive or, in other words, the spreading of these liquids should occur upon such solids when the solid surfaces are *pure*.

The matter of attributing the nonspreading of liquids on some solid surfaces to the presence of impurities on the surface has been questioned by Bulkley and Snyder.[3] They agree with Harkins and Feldman that, on the basis of energy considerations alone, organic liquids should spread on clean solid surfaces. However, they point out that the apparent anomaly in the behavior of fatty oils and fatty acids on highly polished, scrupulously cleaned metal surfaces cannot be explained on the basis of Harkins' and Feldman's considerations. Thus, mineral oils spread completely on such metal surfaces, whereas fatty liquids which are possessed of superior spreading tendency by Harkins' and Feldman's criteria, spread on such metal surfaces hardly at all.

In order to establish the part played in laundering by the spreading coefficient, it is necessary to review the conditions under which spreading is to be expected to take place. Fabrics are composed of

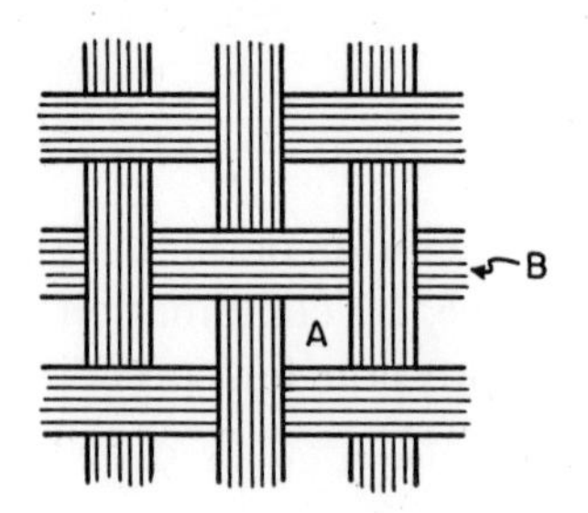

Figure 5–1. Illustration of capillary systems in fabric.
A—Capillaries between threads.
B—Capillaries between fibers of each thread.
A third possible system is the capillaries between the fibrils of each fiber (see Chapter 10).

two and possibly three capillary systems as indicated in Figure 5–1. An additional capillary system may exist as the spaces between adjacent pieces of fabric, particularly when several articles are bunched together in a wash net. Now, to completely saturate the fabric with wash solution, two means of penetration are involved. The first, immersional penetration, depends only on gravitational forces for the extent to which it takes place. Those interstices which are sufficiently large may be penetrated by the wash solution without any dependency on capillary or surface activity. The second means of penetration brings true capillary spreading action into play and it is only by this means that the multitude of fine capillary systems in the wash load can be penetrated. The mass of the wash solution is not in itself sufficient to effect complete penetration into the fine capillaries. It is here where spreading coefficient, as a measure of wetting by spreading, assumes greatest significance. Thus, as the space to be penetrated becomes finer, we have a greater dependency on surface activity.

[3] Bulkley, R., and Snyder, G. H. S., *J. Am. Chem. Soc.*, **55**, 194–208 (1933).

When there is no contact angle between a liquid and a solid (the solid being the surface of a capillary) the maximum penetration of a capillary by the liquid is realized. For the case of a horizontal capillary and a positive contact angle, Washburn[4] has derived the equation

$$x^2 = \left(\frac{\gamma_L \cos \theta}{2m}\right) rt \tag{6}$$

where x is the distance traveled in time t, r is the radius of the capillary, and m the viscosity of the penetrating liquid. Powney[5] has pointed out that this equation fits only when the capillary system is well defined and when the surface tension and viscosity of the liquid do not change during the process of penetration. In the case of pene-

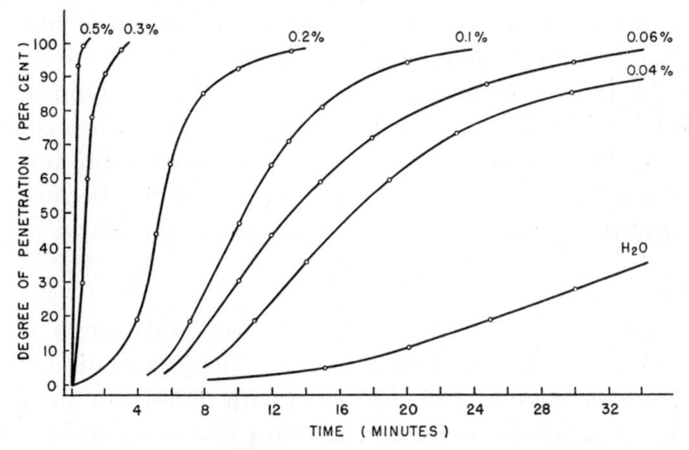

Figure 5–2. Rates of penetration into cotton fabric for sodium oleate solutions at various concentrations.[5]

tration of the capillaries of a fabric by detergent solution, Powney indicates that, in addition to Washburn's idealized conditions, penetration by the solution is dependent on displacement and release of entrapped air and on changes in surface tension and viscosity of the detergent solution by reason of selective adsorption by the fibers.

Powney has devised a means for measuring penetration, based on electrical conductivity through 5 plies of the fabric in question. His results with cotton fabric and sodium oleate solutions are shown in Figure 5–2, in comparison with data for water and cotton. It is of

[4] Washburn, E. W., *Phys. Rev.*, **17**, 273–83 (1921).

[5] Powney, J., "Wetting and Detergency Symposium," pp. 185–96, Brit. Sect. Intern. Soc. Leather Trades' Chem., 1937.

particular interest to note that, at an oleate concentration just above 0.1 per cent, the rate of penetration becomes very rapid. This point corresponds to the concentration for maximum surface activity of the solution, as determined by other means. Powney considers that the further increase in rate of penetration at concentrations above about 0.1 per cent, even though surface tension is also increasing above this point, may be due to spreading of the solution along the walls of the capillaries in advance of the main column of solution.

A direct method for obtaining a quantitative measure of the rate of wetting of a fabric by a detergent solution is that employing the canvas disc wetting test. The "wetting time" is taken as the time in seconds required for a standard canvas disc to become wetted to the extent that it will settle through the surface of the detergent solution. On the basis of this technique, Dreger and coworkers[6] have reported the wetting times shown in Figure 5–3 for several sodium alkyl sulfates. For the identification of these compounds reference is made to Table 4–1, Chapter 4. The technique of the canvas disc test cannot be considered as strictly parallel to actual washing because of the fact that wetting proceeds only from one side of the fabric.

Foaming

A foam may be considered as a gas-in-liquid system in which bubbles of gas are dispersed more or less stably in a continuous liquid phase. In detergency we are concerned with bubbles of air dispersed in detergent solutions or incased in films of the detergent solution and will limit our discussion to those well-known gas-in-liquid systems.

An air bubble rising through pure water is incased in an interfacial "skin of tension" as evidenced by the spherical shape of the bubble. The composition of the water at the bubble interface and at the surface being the same as that in the body of the water, it is to be expected that, when the bubble leaves the support of the water and emerges at the surface, it will burst because of the high surface tension of its "skin" or incasing film in relation to the low elasticity. An air bubble rising through a detergent solution is also incased in an interfacial "skin" but here the "skin" is more than simply one of surface forces. The detergent has concentrated in the interface between the air of the bubble and the liquid because of positive adsorption, so

[6] Dreger, E. E., Keim, G. I., Miles, G. D., Shedlovsky, L., and Ross, J., *Ind. Eng. Chem.*, **36**, 610–7 (1944).

that now the "skin" is in reality a third physical phase in the system. Upon contacting the surface of the solution, the bubble encounters the surface film (of a composition comparable to that of the bubble

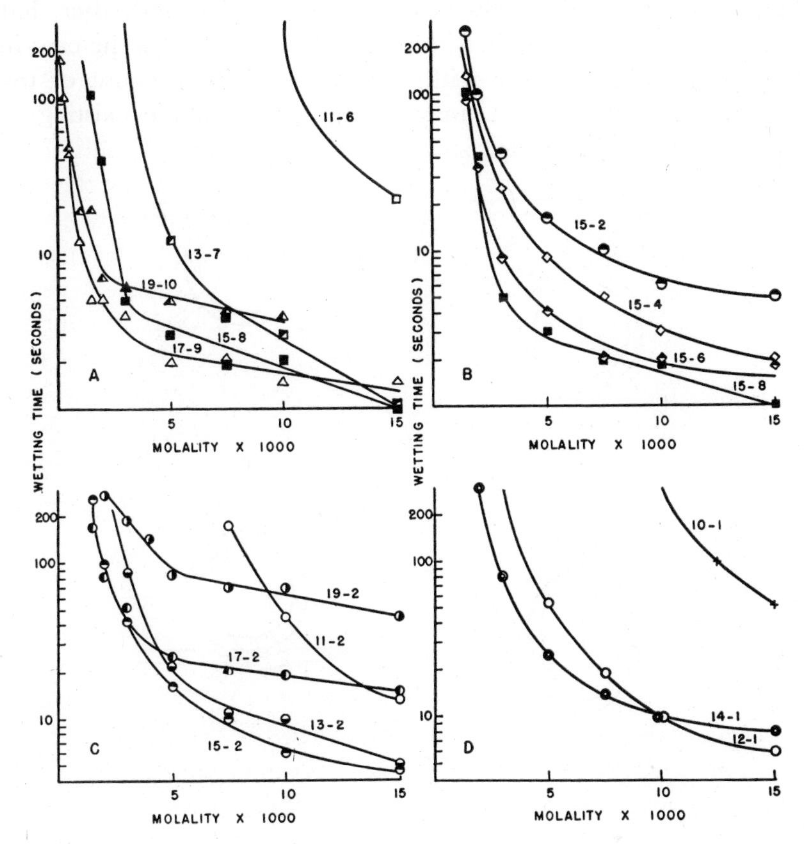

Figure 5–3. Canvas disc wetting time vs. concentration of various sodium sulfates at 43.3–46.7°.[6]

A—Sodium *sym-sec*-alcohol sulfates.
B—Isomeric sodium *sec*-pentadecanol sulfates.
C—Sodium *sec*-alcohol sulfates.
D—Sodium *n*-alcohol sulfates.

(See Table 4–1, Chapter 4, for identification of compounds.)

film) and acts as a spherically shaped wedge driving upward to distort the surface film. The fact that the bubble film does not at this point merge with the surface film and thus disappear is held by Foulk[7] to be due to the presence of a layer of solution between the two films,

[7] Foulk, C. W., *Ind. Eng. Chem.*, **21**, 815–7 (1929).

perhaps as illustrated in Figure 5–4. As this layer is not of the same composition as the films, there results a definite resistance to mixing of the two films. The final bubble film upon emerging on the solution surface now has considerably lower surface tension and, thus, less tendency to compress the air within the bubble than was the case in pure water. This, together with greater elasticity because of the detergent molecules present, may result in the bubble existing as foam for an appreciable time.

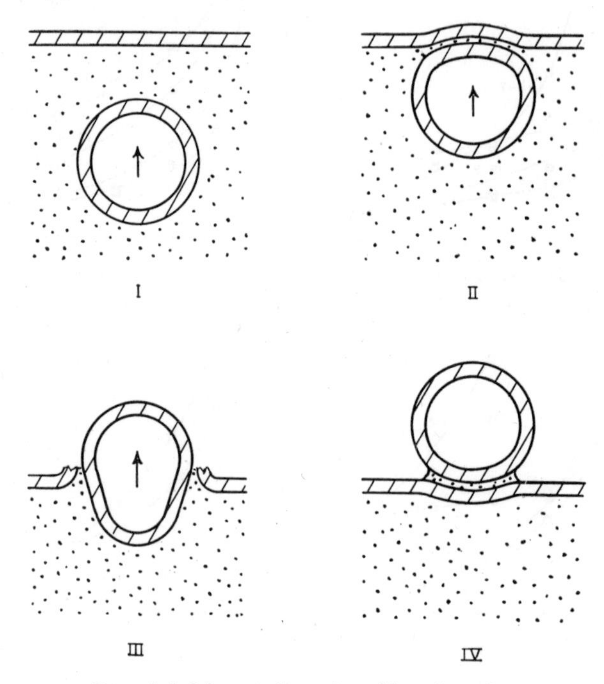

Figure 5–4. Schematic illustration of foam formation.

True foam formation from detergent solutions results when many air bubbles, each with its individual interfacial "skin," emerge from the water side by side and one under the other. Foam then consists of gas bubbles of various sizes, incased in thin layers or laminae of liquid; however, the individual bubbles are no longer spherical in shape because of the influence of adjacent bubbles.

Foams of interest in laundering are only those which persist for some time, as for at least several minutes. Fugitive foams or bubbles which persist only for a few seconds are of no significance. Therefore, our discussion will now be directed to the conditions which govern the formation and stability of persistent foams. In the first place, pure liquids do not foam in the absence of a third component. Foulk reviews the following general facts about foam: (1) pure liquids do not foam; (2) a third component must be present in a liquid-gas system to make it foam and nearly all substances induce foaming to widely varying extents; (3) it does not matter whether the added substance is positively or negatively adsorbed in the bubble films. Berkman and Egloff[8] point out that *pure* liquids with low surface tension are close to their boiling points and have a high vapor pressure, from which it follows that thin laminae from such liquids will evaporate too rapidly to form stable bubble films. Low surface tension and low vapor pressure when present together, as when induced by a third, surface-active, component, serve as favorable conditions for the promotion of the foaming process.

In 1924, Bartsch[9] reported an extensive study of the foam-promoting powers of a number of substances, from which he formulated the following criteria for foam formation:

(a) Surface activity, solubility, degree of dispersion, and viscosity are factors in the effectiveness of each substance;

(b) In all dilute solutions, the foam-forming power increases with increasing surface activity (lowering of surface tension);

(c) In concentrated solutions, the influences of dispersion, solubility, and viscosity mask that of surface activity;

(d) Molecular association or colloidal aggregation decreases the foam-forming power of surface-active substances;

(e) High viscosity may not aid in the formation of foam but increases its stability;

(f) Mixed solutions of surface-active substances have greater foam-forming power than the solution of either component, unless mixing lowers the solubility;

(g) The ability to form foam is conditioned by a sharp difference between the concentration in the liquid-air interfacial layer and in the body of the liquid;

(h) In a three-phase foam system (foam-promoting substance a solid) the characteristics of the system are primarily dependent on the concentration of the promoter, but the dispersion and nature of the solid phase play important roles.

---

[8] Berkman, S., and Egloff, G., *Chem. Rev.*, **15**, 377–424 (1934).
[9] Bartsch, O., *Kolloidchem. Beihefte*, **20**, 1–49 (1924).

Ostwald and Steiner[10] question the role of surface tension in foaming, to which Bartsch[11] replies that foam stability increases up to the point where surface-tension lowering is practically a maximum, and then decreases rapidly, the foaming power being lost if the surface-active substance is added in excess. He considers the point of maximum foaming power to be that at which equivalent amounts of solvent and solute (water and soap in the case of soap solutions) are present in the surface layer.

The basis for the explanation of the influence of solubility on the foam-promoting power of a given soap was advanced by Traube and Klein[12] in their statement that only the molecularly dispersed portion of a surface-active agent is related to surface activity. On the other hand, surface activity increases in ascending a homologous series, such as of soaps, while solubility decreases. The decrease in solubility limits the increase in surface activity. Therefore, in a homologous series of soaps, those soaps with the most advantageous relationship between solubility and surface activity, other factors being constant, will show the greater foam-promoting power.

Foam promoters may be present as molecularly or colloidally dispersed electrolyte, as a protective colloid, or a solid promoter. Soap may be considered to fit the first two classes and solid soil from wash work to fit the latter class. Thus, in laundering, all three classes and the manner in which they form and stabilize foams must be considered.

In comparing the foaming characteristics of different soaps, consideration must be given both to profuseness and to stability of the foam. Several such comparisons[13-20] permit the following summarization:

(a) The foaming power of soaps from several natural fats and

---

[10] Ostwald, W., and Steiner, A., *Kolloid-Z.*, **36,** 342–51 (1925).

[11] Bartsch, O., *Kolloid-Z.*, **38,** 177–9 (1926).

[12] Traube, I., and Klein, P., *Kolloid-Z.*, **29,** 236–46 (1921).

[13] Anon., *Soap,* **1,** No. 10, 19–21 (1926).

[14] Petrova, N. N., Komarova, M. I., and Bobuileva, E. N., "Vsesoyuznuiĭ Nauch.-Issledovatel," *Inst. Zhirov, Untersuchungen uber Physikochemie der Waschwirkung,* **1935,** 93–115.

[15] Smirnov, N. M., "Vsesoyuznuiĭ Nauch.-Issledovatel," *Inst. Zhirov, Untersuchungen uber Physikochemie der Waschwirkung,* **1935,** 116–25.

[16] Smith, P. I., *Am. Perfumer,* **43,** No. 1, 55, 57 (1941).

[17] Miles, G. D., and Ross, J., *J. Phys. Chem.*, **48,** 280-90 (1944).

[18] Merrill, R. C., Jr., and Moffett, F. T., *Oil & Soap,* **21,** 170–5 (1944).

[19] Mikumo, J., *J. Soc. Chem. Ind. Japan,* **37,** Suppl. binding, 591–3 (1934).

[20] Mikumo, J., *J. Soc. Chem. Ind.*, **52,** 65–68T (1933).

oils descends in approximately the following order: palm oil; palm kernel oil or tallow; coconut oil.

(b) The foaming power and foam stability of solutions of single saturated soaps increase with increase in molecular weight, sodium stearate being a firm-lathering soap (if the temperature is sufficiently high to assure adequate solubility) and sodium laurate being an example of a profusely foaming soap with a nonpersistent foam.

(c) Highly unsaturated single soaps, such as sodium linoleate, yield poorly foaming solutions.

(d) The stability of foams of solutions of a homologous series ot soaps at a given concentration, when each solution is adjusted to the optimum pH for the particular soap, increases with increasing chain length of the soap molecule (see also Chapter 9, page 169).

(e) The stabilities of foams from solutions of soaps of most of the natural fats and oils (solutions of naturally *mixed* soaps) are greater than those of foams from solutions of single soaps, unless one or more of the soaps in the mixture is of such high molecular weight as to be definitely insoluble at the particular temperature; this accounts for the inability to correlate summary (a), above, and the composition of the various naturally mixed soaps (see Table 2–1, Chapter 2) with summaries (b), (c), and (d).

The effects of composition and concentration on foaming power and foam stability of a series of saturated and a series of unsaturated sodium soaps are illustrated in Figure 9–8, Chapter 9, taken from the work of Miles and Ross.[17] A similar study of sodium alcohol sulfates has been reported by Dreger and coworkers,[6] as shown in Figure 5–5. (See Table 4–1, Chapter 4, for identification of the sulfates shown on the figure.)

Berkman and Egloff,[8] in an extensive review of the physical chemistry of foams, point out the importance of heterogeneity of the liquid-air boundary film. In the two cases, first, when the concentration of surface-active agent is so small that the boundary layer is composed essentially of water molecules and second, when the concentration of the surface-active agent is at saturation so that the boundary layer is composed essentially of molecules of the agent, foaming is not induced. In all intermediate cases, in which the boundary layer is made up of two or more types of molecules and is therefore heterogeneous, foaming occurs. Thus, foaming ability increases with

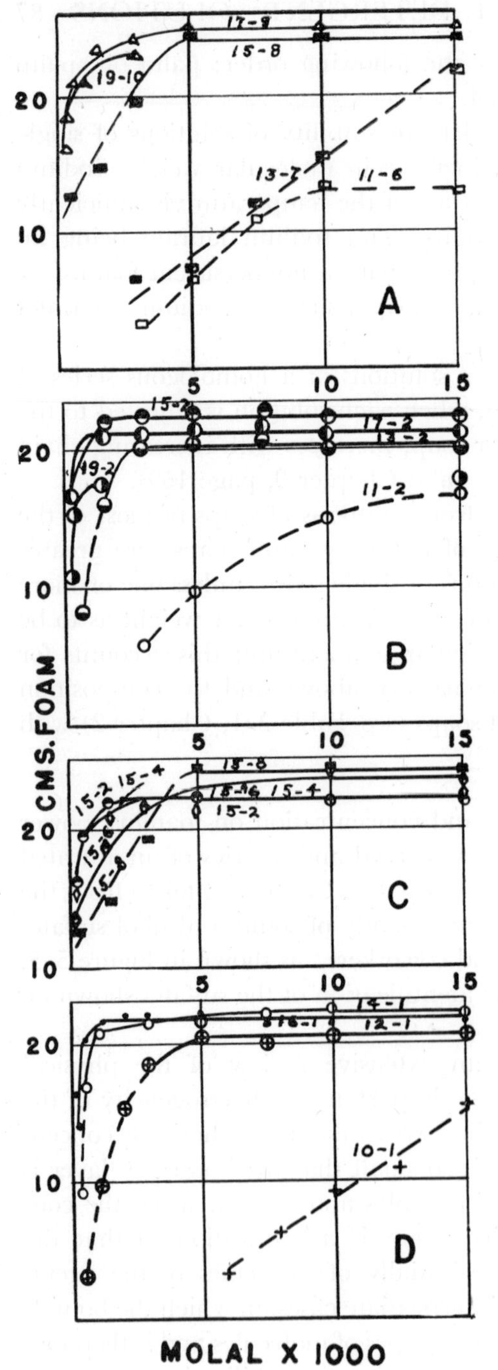

Figure 5–5. Centimeters of foam vs. concentration of various sodium sulfates at 46°C.[6]

A—Sodium *sym-sec*-alcohol sulfates.
B—Sodium *sec*-alcohol sulfates.
C—Isomeric sodium *sec*-pentadecanol sulfates.
D—Sodium *n*-alcohol sulfates.

(See Table 4–1, Chapter 4, for identification of compounds.)

increased heterogeneity in the composition of the boundary film. Any change in prevailing conditions in the system, such as pH, temperature, concentration, and addition of a second surface-active agent, which changes the heterogeneity of the boundary film, will correspondingly change the foaming characteristics of the system.

As has been demonstrated by Schütz[21] the surface tensions of various phases of foaming solutions may have the relation

$$R > O > F \tag{7}$$

wherein F is the surface tension of the liquid formed by separating off and settling the foam, O is the surface tension of the original solution, and R is the surface tension of the residual solution. Starting with a solution of a surface-active agent such as soap, this relationship demonstrates the positive adsorption of the agent in the bubble films. Mikumo[19, 20] shows that this action of adsorption may be preferential. Thus, in the case of two soaps in the same solution, he finds the soap foam to be richer in the soap of higher molecular weight.

Preferential adsorption in foam films leads to a basic and important property of soap foams and other foams, that of *flotation* of solid material. The adsorptive action of liquid films extends to finely dispersed solid particles. This is evidenced from the previously discussed role of finely dispersed solids as foaming agents. Flotation is based on the preferential adsorption of solid material in bubble or surface films and depends on selective wetting whereby the surface film will wet and hold the solid particles more strongly than will the main body of the solution. Solid particles adsorbed in the liquid-air interface of an air bubble will rise with the bubble and be held in the foam at the surface of the liquid.

Emulsification

Emulsions are systems consisting of one liquid, the dispersed phase, dispersed as droplets in another liquid, the continuous phase. The two most common types of liquids in emulsions, and the only ones with which we are concerned in considering emulsification in laundering, are water and oil. The water phase may be an aqueous detergent solution. The oil may be any of several of the liquid oil soils encountered in laundering.

---

[21] Schütz, F., *Nature*, **139**, 629–30 (1937).

Emulsions of water and oil are of two main types, namely, the oil-in-water type wherein the oil is dispersed as droplets in the water, and the water-in-oil type wherein the water is dispersed as droplets in the oil. Here again our interest is, for the most part, limited, this time to the oil-in-water type of emulsions, as the conditions generally prevailing in laundering are not conducive to the water-in-oil type. However, this latter type must be considered to some extent because of the possibility of reversion from one type to the other under certain conditions, even over a wide range of variation in proportion of the two liquids.

The main criterion for the formation of emulsions, at least for *stable* liquid-in-liquid dispersions, is that a suitable third phase be present. This third phase, the emulsifying agent, may be any of several substances, liquid or solid. The agents we are primarily interested in are soaps and synthetic detergents, together with certain solid materials.

Dispersion of one liquid as droplets in another liquid requires that the two be no more than slightly miscible. The dispersion is attended by an enormous increase in interfacial area with corresponding increase in total surface energy. The mechanical work done to effect the dispersion is thus converted to additional free surface or interfacial energy. As pointed out by Clayton,[22] the specific surface, i.e., surface/volume, of the dispersed liquid must be increased at least 10,000 times the original bulk value for noticeable effects to come into play. In emulsions, where it is not uncommon for the dispersed liquid to be in droplets as small as 10^{-5} cm. in diameter, surfaces exceeding 600,000 sq. cm. per cc. may be found. Purely as an illustration of the change in free surface energy which might attend such dispersion in air, the following calculation is shown:

Starting with one spherical droplet, one cc. in volume, of a liquid having a surface tension of 25 dynes per cm., the total free surface energy of the droplet will be 25×1.24^2 (diameter) $\times 3.1416 = 121$ ergs. Dispersion of this initial droplet of liquid to droplets of a uniform diameter of 10^{-5} cm. would result in an increase in the total surface area to 600,000 sq. cm. and a corresponding increase in total free surface energy from 121 ergs to 15,000,000 ergs (assuming no change in surface tension for the small droplets).

---

[22] Clayton, W., *J. Soc. Chem. Ind.*, **38**, 113–8T (1919).

In the case of two liquids, one to be dispersed in the other, the work required to effect the dispersion of the one liquid may be somewhat less, particularly if the surface tensions of the two liquids are closely alike and the interfacial tension is consequently low. In a two-component system such as oil-in-water, however, the work of dispersion is still high because of the high interfacial tension between the two phases.

The first function of soaps and synthetic detergents in the action of emulsification may now be introduced. These agents which, because of their surface activity as previously discussed, lower the interfacial tension between oil and water are directly capable of reducing the amount of work necessary to attain a particular degree of dispersion in an oil-in-water system. Under ordinary conditions, soaps and the like do not produce the action of emulsification but, due to their effect on interfacial tension, they greatly facilitate dispersion by applied mechanical agitation. Spontaneous emulsification without mechanical agitation is a function of alkalies rather than of preformed detergents and will be discussed in the next chapter (page 119).

It has been recognized that lowering of interfacial tension is not the only factor contributing to ease of emulsification. Thus, Pickering[23] was probably the first to study the efficacy of certain insoluble solids as emulsifying agents. From such work came the realization that agents having no appreciable action on interfacial tension can be effective emulsifying agents, particularly with respect to greater stability.

Assuming the oil to have been dispersed as droplets in a detergent solution by means of suitable mechanical agitation aided by the decreases in surface energy effected by the detergent, the next and perhaps the more critical problem is the stabilization of the emulsion. Stabilization requires prevention of coalescence of the droplets of oil and also prevention of reversion from the oil-in-water to the water-in-oil type.

An emulsion, to remain stable, must repel several forces or factors, as follows:

(a) The gravitational effect of difference in specific gravities of the oil phase and the water phase;

---

[23] Pickering, S. U., *J. Chem. Soc.*, **91**, 2001–21 (1907).

(b) The force or attraction of coalescence as is to be expected between, say, two "bare" droplets of the same liquid in contact with each other, particularly when in the presence of another liquid as different in character as water; this force is a manifestation of the high total free surface energy of the system, the tendency to coalesce being the natural tendency for many small droplets to reduce their surface and, thus, surface energy by reverting to the original "bulk" state;

(c) The adsorbing action for the oil droplets by a third phase, such as fabric immersed in the emulsion.

The gravitational effect on stability of emulsions may be counteracted by reducing the difference in specific gravity between the dispersed and continuous phases, by increasing the viscosity of either the continuous phase or the entire emulsion, or by continuation of mechanical agitation. Under the conditions present in laundering, neither the difference in specific gravity between the two phases nor the viscosities of the continuous phase and of the emulsion can be varied at will. Therefore, continued mechanical agitation must be depended on for overcoming the gravitational effect.

The problem of overcoming coalescence between droplets is not so simple. The stability of emulsions against coalescence is considered to depend on several factors:

(a) Each droplet of oil carries an electrical charge, the origin of which may be loss of ions from the surface of the droplet or adsorption of ions from the continuous phase. The sign of the charge of oil droplets in water is negative, as the oil has a lower dielectric constant than the water. It follows then that the electrical repulsion between such droplets of like charge aids in preventing coalescence.

(b) There exists between oil droplets and the continuous phase a definite potential difference or contact potential which is of a magnitude in the range of -0.05 volt and which, according to Ellis,[24] is independent of the kind of oil used or even the degree of purity of any given oil. The stability of the emulsion decreases as the potential difference between its two phases is decreased, such as by the addition of electrolyte. Powis[25] has shown that there is a critical potential, -0.030 volt, at which an emulsion is most stable. All anions in an

[24] Ellis, R., *Z. physik. Chem.*, **78**, 321–52 (1912).
[25] Powis, F., *Z. physik. Chem.*, **89**, 91–110 (1914).

aqueous solution in which the oil droplets are dispersed exhibit a tendency to make the potential difference negative, and all cations a tendency to make it positive. The influence of both anions and cations increases considerably with their valence. (For further details of such electrical phenomena, see Chapter 7.)

(c) Bancroft[26] considers that, in many cases, the existence of an emulsion depends on the formation of a film or membrane around each droplet. Such a film may be formed by the positive adsorption of detergent in the interface around the droplets, or even by the adsorption of particles of certain finely dispersed solid material in the interface. Provided the concentration of detergent in the continuous phase at the start is sufficiently high it is possible for the adsorption of detergent in the interface to continue to the point where the resulting film is practically one of liquid or semi-solid detergent. The emulsifying agent may then be considered almost to form a rather impermeable protective "skin" around each droplet to prevent or inhibit coalescence. A further factor in stabilization of a dispersed oil droplet, which may be in reality only an extension of Bancroft's idea of a protective "skin," is the protective action of the atmosphere of water molecules held around the droplet by solvation to inhibit direct droplet-to-droplet contact.

(d) Clayton[22] points out another property possessed by droplets in an emulsion which aids against both coalescence and the gravitational effect, namely, that of Brownian movement or pedesis. For such action to be effective, the diameter of the droplet should not exceed about 0.004 cm.

A special class of emulsifiers consists of the solid agents. Even though rarely added as such, solid emulsifying agents probably have some significance in laundering. Thus, finely dispersed solid soil, after removal from the fabric, may become adsorbed at an oil-water interface and exert a stabilizing action. In such a case the tendency is for that liquid which more readily wets the solid to become the continuous phase. Such solid emulsifiers are not surface active in the sense that they exert any appreciable effect on surface or interfacial tension. However, they are surface-active in the sense that, at an interface between two liquids, the relative extents to which the two liquids wet the solid determine its efficacy as an emulsifier. As has

---

[26] Bancroft, W. D., *J. Phys. Chem.*, **16**, 177–233 (1912).

been pointed out by Fain and Snell,[27] only those solid particles which fall in Class (*c*) proposed by Reinders[28] (see page 69) may serve as emulsifying agents.

Other factors being equal, the continuous phase of a stable emulsion is that phase which is the stronger solvent for the emulsifying agent. Likewise, the relative proportions of the two phases tend to favor one as the continuous phase, the tendency being for the most abundant phase to be the continuous one. Thus, in laundering where the ratio of water or detergent solution to oily soil is very high, the preferential solubility of the detergent in the water and the preponderance of amount of detergent solution favor the formation of oil-in-water dispersions.

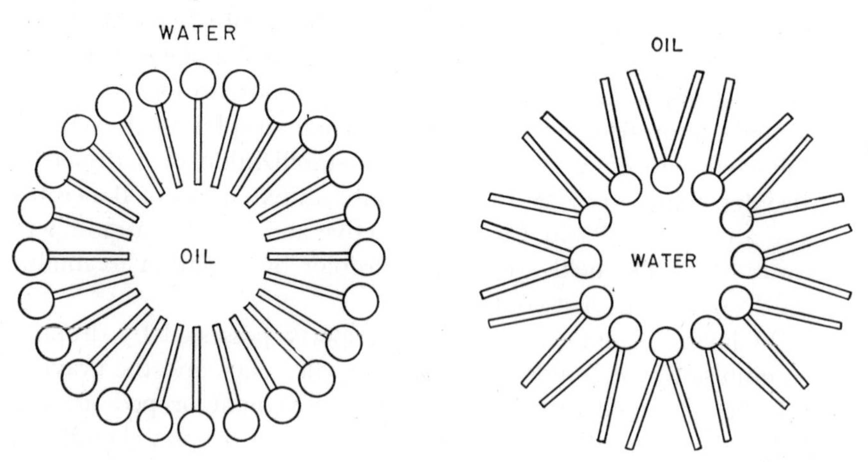

Figure 5–6. Illustration of the effects of relative volume of the polar group on type of emulsion and size of droplets.

Based on Langmuir's and Harkins' theories of the orientation of molecules in interfacial films, Finkle, Draper, and Hildebrand[29] point out that the relative volume of the polar group of the detergent molecule, with respect to that of the fatty end of the molecule, should play an important part in determining the type and stability of the emulsion, as indicated in Figure 5–6. Thus, for example, due to the width of the polar group in a sodium soap being greater than the

---

[27] Fain, J. M., and Snell, F. D., *Ind. Eng. Chem.*, **31**, 48–51 (1939).

[28] Reinders, W. von, *Kolloid-Z.*, **13**, 235–41 (1913).

[29] Finkle, P., Draper, H. D., and Hildebrand, J. H., *J. Am. Chem. Soc.*, **45**, 2780–8 (1923).

width of the fatty chain, the curvature of the film of soap adsorbed at the interface is convex toward the water. Further, the geometry of this curvature, as determined by these relative dimensions, governs the ultimate fineness of dispersion of the enclosed droplets. In the case of soaps of polyvalent metals, the curvature of the film with respect to water is reversed because of the presence of two or more fatty chains in one soap molecule, and a water-in-oil type of dispersion results. This leads to the discussion of another type of instability that might arise under conditions of laundering, namely, reversion of type of emulsion. Assume that we start with a sodium soap and have attained a dispersion of oily soil in the soap solution. In the presence of calcium or magnesium ions that might be present in the solution as "hardness" in the water, the sodium soap in the stabilizing film may be converted to calcium or magnesium soap and the tendency will be for reversion of the original oil-in-water dispersion to a water-in-oil dispersion. However, under the prevailing conditions of relative concentration, the final product probably will be a flocculant mass of oil, calcium or magnesium soap, and water.

Peptization

Some controversy exists as to the proper definition of the term "peptization." Perhaps in a strictly technical sense the action of peptization pertains only to the transformation of a gel into a sol or, in other words, the liquefaction of a gel by the addition of a small quantity of a dispersing or peptizing agent in the absence of mechanical or electrical agitation. Slater[30] has presented an extensive discussion of the various uses and misuses of the term. However, in the present discussion, for lack of a better term, "peptization" will be considered as the spontaneous dispersion of agglomerated particles by a dispersing agent, whether the particles are agglomerated as clumps of solid material or as gels and whether the individual particles are of true colloidal size or only approaching colloidal size. As such, the term will not include applied mechanical or other physical action that generally attends disintegration and dispersion. Specifically, the present discussion will be limited to the peptizing action of soaps and synthetic detergents on agglomerates of discrete particles of solid soil and on semi-solid albuminous or gelatinous material.

---

[30] Slater, A. V., *J. Soc. Chem. Ind.*, **44**, 499–506T (1925).

The literature is practically void of references pertaining to soaps and synthetic detergents as peptizing agents in the sense defined above, although much attention has been paid to these agents as deflocculating or stabilizing agents. To keep the discussion of peptization properly separate from that of stabilization, it is, therefore, necessary to resort to analogous cases wherein the peptizer is some material other than a detergent, and to hypotheses as to the action of detergents as peptizers of solid soil aggregates.

Bancroft[31] considers that the lowering of the surface tension of a material by the adsorption of a second material will tend to disintegrate or to peptize the adsorbing phase. Thus, it would follow that a clump or aggregate of small solid particles, upon adsorption of a surface-active agent on its surface and within its interstices, would tend to peptize because of the lowered surface tension. However, Slater points out that, although peptization is the result of adsorption, so also is flocculation, the antithesis of peptization. Considering peptization only in the strict sense of dispersion of gels or other aggregates of colloid, Slater considers the following to be the criteria:

(a) In most cases, one of the ions of the colloid must be present in the peptizer or be introduced by secondary reaction;

(b) The peptizer should be capable of reaction with the colloid, although this reaction may not necessarily be in true stoichiometric relation (as, for example, the peptization of a flocculant precipitate of ferric hydroxide hydrogel by ferric chloride or indirectly by hydrochloric acid).

He further considers that increased electric charge (as for effecting deflocculation) and lowered interfacial tension are not sufficient in themselves to cause actual dispersion. Some other force is necessary, namely, true solution pressure. In accordance with this theory, as originally advanced by Ruff,[32] a peptizer or a group from the peptizer is attached by surface valency to the surfaces of a colloid aggregation. The surfaces thus become the seat of still more active surface valencies, the attraction (solution pressure) of which for the molecules of the dispersion medium leads to "pulling apart" of the aggregation into discrete particles.

---

[31] Bancroft, W. D., *J. Phys. Chem.*, **20,** 85–117 (1916).
[32] Ruff, O., *Kolloid-Z.*, **30,** 356–64 (1922).

A basic principle for deflocculation, based on work by Fuchs,[33] may be borrowed for the present discussion because of its apparent direct applicability to peptization; namely, deflocculation in water (and, likewise, peptization in water) becomes nearer perfect the more nearly do the three factors—attraction between the discrete particles, attraction between each particle and water, and attraction of water for itself—approach the relation $A_{pp} < A_{pw} = A_{ww}$. Chapin[34] shows that an effective deflocculant (and, presumably, also an effective peptizer) must accordingly possess a high attraction for water in order that it may impart to the particles upon which it is adsorbed a similar high attraction for water; however, such attraction must be counterbalanced by positive adsorption of the peptizer upon the surfaces of the particles. Harkins, Davies, and Clark[35] have shown that these two contradictory requirements can be met only by highly unsymmetrical molecules of pronounced polar-nonpolar structure which build an oriented layer at the interface between particle and water. Soaps and synthetic detergents meet this dual requirement, and certainly their pronounced deflocculating powers would indicate parallel pronounced efficacy as peptizers.

Much of the discussion of formation and stabilization of dispersed systems, as will appear later, will be directly applicable to peptization, and the reader is referred to those discussions for further details.

Solubilization

It has long been known that aqueous solutions of soaps and of many other substances, wherein the solute is present at least partially as colloidal micelles, are capable of "dissolving" considerable amounts of liquid or solid material which is insoluble in water alone. Perhaps the earliest orderly study of this phenomenon is that of Engler and Dieckhoff,[36] in 1892, based on numerous measurements of the dissolving power of soap solutions of relatively high concentration for hydrocarbons, fatty acids, phenols, and mixtures. Table 5–1 is borrowed from their data, as republished recently by McBain.[37] They found that solutions in rosin soap clouded on warming and cleared again on

---

[33] Fuchs, *Exner's Repert. Physik.*, **25**, 735 (1889).

[34] Chapin, R. M., *Ind. Eng. Chem.*, **17**, 1187–91 (1925).

[35] Harkins, W. D., Davies, E. C. H., and Clark, G. L., *J. Am. Chem. Soc.*, **39**, 541–96 (1917).

[36] Engler, C., and Dieckhoff, E., *Arch. Pharm.*, **230**, 561 (1892).

cooling; e.g., heating cold 25 per cent sodium rosin soap saturated with 32 cc of turpentine to 100°C separated 20 cc of the oil, which completely dissolved again on cooling. This constituted the first evidence of the thermodynamic reversibility of the phenomenon. Sisley[39] reported experiments with dyes to show that emulsions have

TABLE 5–1. SOLUBILITY OF HYDROCARBONS IN ONE HUNDRED CC. SOAP SOLUTION AT ROOM TEMPERATURE [36, 37]

	Benzene, cc	Toluene, cc	Xylene, cc	Turpentine, cc
Na stearate 10%[a]	1.6	1.5	1.0	0.8
Na palmitate 10%[a]	1.8	1.3	1.4	0.4
Na oleate 10%	10	9.6	7.4	7
Na rosin soap 10%	5.2	4.4	3.6	5.8
Na rosin soap 15%	8.8	8.2	8.0	11.2
K rosin soap 15%	8.4	8.0	6.8	9.0
Na rosin soap 25%	20	18	17	32
Water only	0.09 [38]	0.05[38]	insol.[38]	insol.

[a] At elevated temperature.

at times a much greater solvent action than the substances from which they are formed.

Smith,[40] in 1932, reported on a study of the solubility of 26 organic liquids in aqueous 10.8 per cent sodium oleate solution, and in all cases found pronounced increase in solubility over that in water alone. He explains this solvent power of the soap solution by postulating absorption of the organic solute in the colloidal soap particles. The possibility of such sorptive action had been previously suggested by McBain and Bolam[41] in connection with a study of the extent of hydrolysis of soaps.

Explanation of the mechanism of solubilization requires reference to the structure of the micelle, as illustrated in Figure 3–13, Chapter 3. Many investigators have studied solubilization by means of x-rays, based on the spacings of this or quite similar structures.

Kiessig and Philippoff[42] report that the spacings of the lamellar

---

[37] McBain, J. W., "Solubilization and Other Factors in Detergent Action," "Advances in Colloid Science," pp. 99–142, New York, Interscience Publishers, Inc., 1942.

[38] "Handbook of Chemistry and Physics," 29th ed., Cleveland, Chemical Rubber Publishing Co., 1945.

[39] Sisley, P., *Bull. soc. chim.*, **21,** 155–7 (1917).

[40] Smith, E. L., *J. Phys. Chem.*, **36,** 1401–18 (1932).

[41] McBain, J. W., and Bolam, T. R., *J. Chem. Soc.*, **113,** 825–32 (1918).

[42] Kiessig, H., and Philippoff, W., *Naturwissenschaften*, **27,** 593–5 (1939).

micelles in solutions of sodium oleate undergo an expansion in thickness when benzene is dissolved therein, the normal spacing increasing from 91 Å to 127 Å, as shown in Figure 5–7. The sheets of benzene are 36 Å thick (at saturation) and those of water 42 Å thick, each of the two separated layers of soap molecules contributing 24.5 Å, making a total of 127 Å for the complete unit.

Stearns and coworkers[43] have recently made an extensive quantitative study of the amounts of various hydrocarbons solubilized in

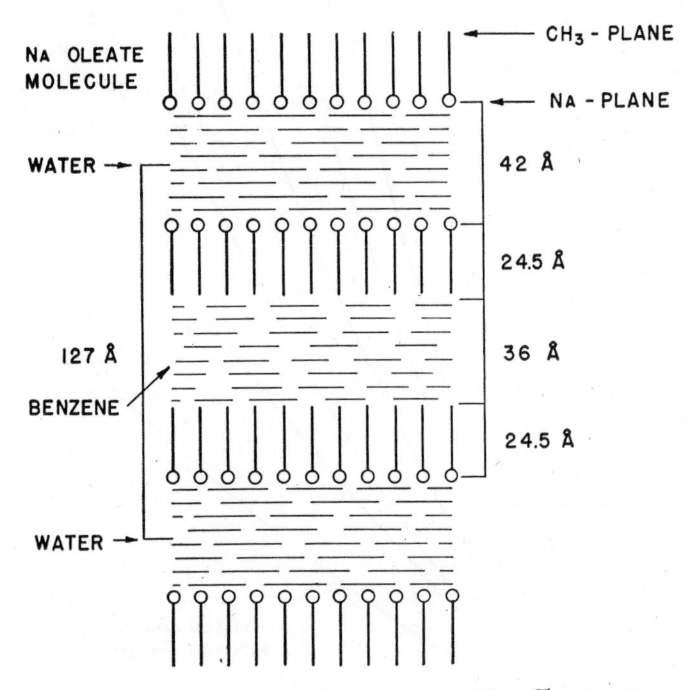

Figure 5–7. Idealized illustration of a lamellar micelle in 9.12 wt. % sodium oleate solution with 0.791 g. benzene per gram of oleate.[37]

solutions of potassium laurate and of potassium myristate. The hydrocarbons used and the "solubilities" thereof in soap solutions of different concentrations are shown in Figure 5–8, taken from their data. They found further that the solubilizing power of the soap solution per unit amount of soap increases as the soap concentration increases; thus, the solubilization *per mol of soap* is greater in a 25 per cent than

[43] Stearns, R. S., Oppenheimer, H., Simon, E., and Harkins, W. D., *J. Chem. Phys.*, **15**, 496–507 (1947).

in a 5 per cent soap solution. For a homologous series of hydrocarbons, the volume of oil solubilized at a constant temperature is, to a first approximation, inversely proportional to the molar volume of the oil. An interesting and valuable confirmation of the existence of a critical concentration for the initial formation of micelles comes from

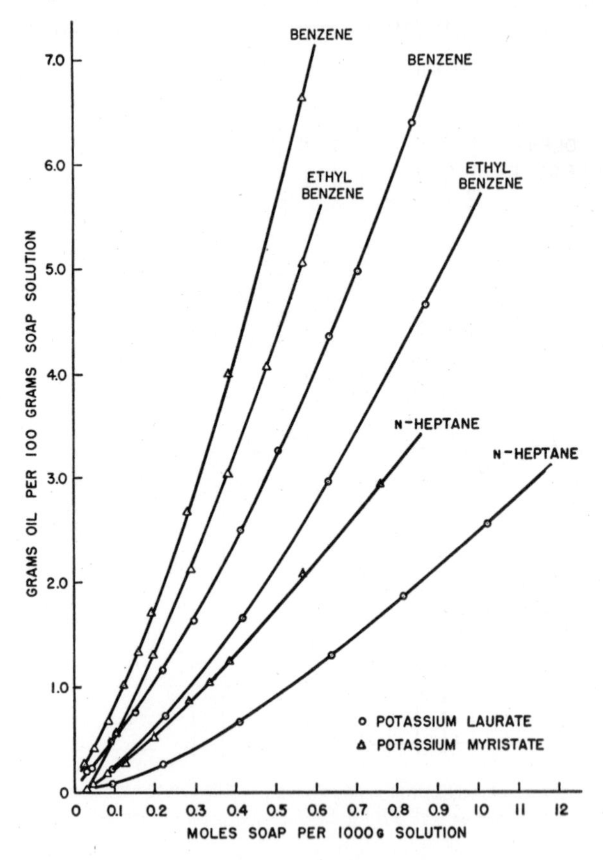

Figure 5–8. Solubilization of hydrocarbons in aqueous solutions of potassium laurate and potassium myristate.[43]

their data, wherein it was found that an increase in the concentration of a soap or other detergent does not increase the solubility of an oil above that in water until a critical soap concentration is attained, corresponding to the critical concentration for formation of micelles.

The ease and extent of solubilization depend to a considerable extent on the composition of the material being solubilized. Thus, as previously reported by McBain and Richards[44] and affirmed by Ralston and Eggenberger,[45] increase in chain length of a hydrocarbon is attended by a drastic reduction in its ease of solubilization and hydrocarbons of high molecular weight are scarcely solubilized.

As pointed out by McBain and coworkers,[46] it seems probable that any material, solid or liquid, may be "dissolved" in any solvent through the presence of a few tenths of a per cent of a suitable solubilizing detergent. However, the efficacy of different soaps and synthetic detergents as solubilizers for a given material varies considerably with composition of the detergent and, likewise, the efficacy of a given detergent should vary with the nature of the soil. Most of our present knowledge of solubilization is so new that little has been done towards a study of the solubilization of actual soils encountered in laundering. Most of the insoluble liquids, the solubilization of which has been studied to date, are seldom encountered in laundering, and studies of the solubilization of solids have been limited to water-insoluble dyes. However, quantitative data on the relative solubilizing powers of various detergents for such materials may well serve as indices of the efficacy of detergents as solubilizers in laundering. Typical data of this kind are given in Table 5–2, taken from the work of McBain and Merrill,[47] and in Figures 5–9 and 5–10, taken

TABLE 5–2. SOLUBILITY OF ORANGE OT DYE IN ONE PER CENT AQUEOUS SOLUTIONS OF PURE AND COMMERCIAL SYNTHETIC DETERGENTS AT 25°C [47]

Detergent	Mg. Dye/ 100 cc soln.	Detergent	Mg. Dye/ 100 cc soln.
Inponol WA	8.3	Nacconol NR	16.1
Igepal C	1.87	Nopco 1440	7.7
Igepal CTA	1.80	Santomerse D	4.86
Lamepon 4C	1.10	Santomerse No. 3	7.0
M.P. 191	1.53	Triton K12	4.72
Morpeltex B	3.80	Triton NE	1.67
Nacconol FSNO	3.8		

from the work of McBain and Green.[48] In connection with the latter data, it is of interest to note that, whereas the hydrocarbon chain in the molecule of soap or in the soap micelle increases from the capryl-

[44] McBain, J. W., and Richards, P. H., *Ind. Eng. Chem.*, **38**, 642–6 (1946).
[45] Ralston, A. W., and Eggenberger, D. N., *J. Am. Chem. Soc.*, **70**, 983–7 (1948).
[46] McBain, J. W., Merrill, R. C., Jr., and Vinograd, J. R., *J. Am. Chem. Soc.*, **63**, 670–6 (1941).
[47] McBain, J. W., and Merrill, R. C., Jr., *Ind. Eng. Chem.*, **34**, 915–9 (1942).
[48] McBain, J. W., and Green, Sister A. A., *J. Am. Chem. Soc.*, **68**, 1731–6 (1946).

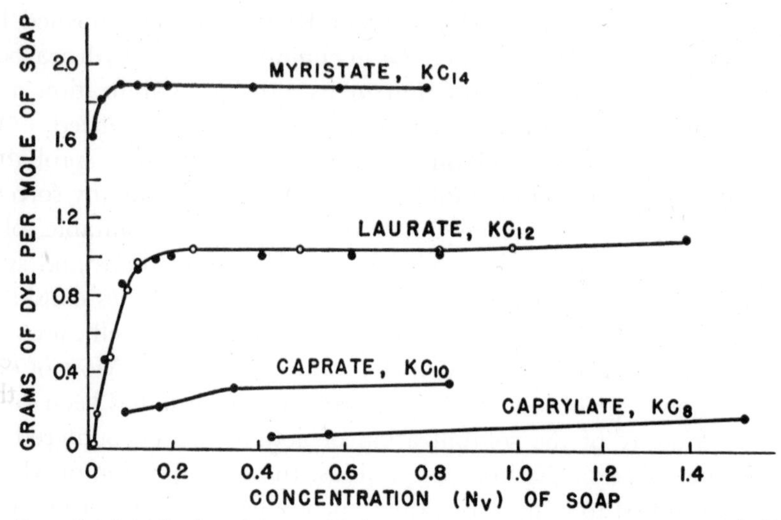

Figure 5–9. Solubilization of Orange OT dye in aqueous potassium soap solutions at 25°C.[48]

ate to the myristate only in the proportion 1 : 1.25 : 1.50 : 1.75, the solubilization increases as 1 : 2.14 : 6.48 : 11.61.

Ion Exchange

As will be discussed in greater detail in Chapter 10, the vegetable fibers such as cotton and linen are primarily cellulosic in nature, and the animal fibers such as wool are primarily proteinaceous.

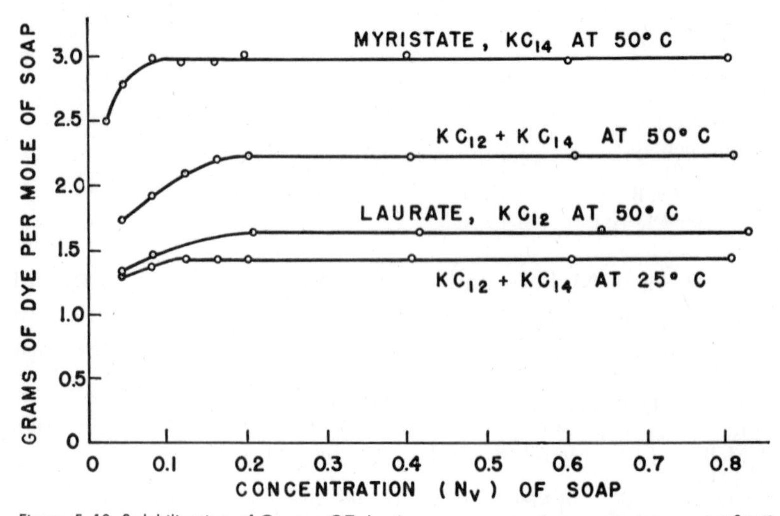

Figure 5–10. Solubilization of Orange OT dye in aqueous potassium soap solutions at 50°C.[48]

Cotton fabric, scoured and bleached, may consist of as much as 99 per cent cellulose. Pure cellulose has the general formula $(C_6H_{10}O_5)_n$, the value of n being as yet indefinite because of lack of information as to the actual length of the complete cellulose molecule. Its chain formula is generally pictured as shown in Figure 5–11. The group enclosed within brackets in the figure is the cellobiose group,

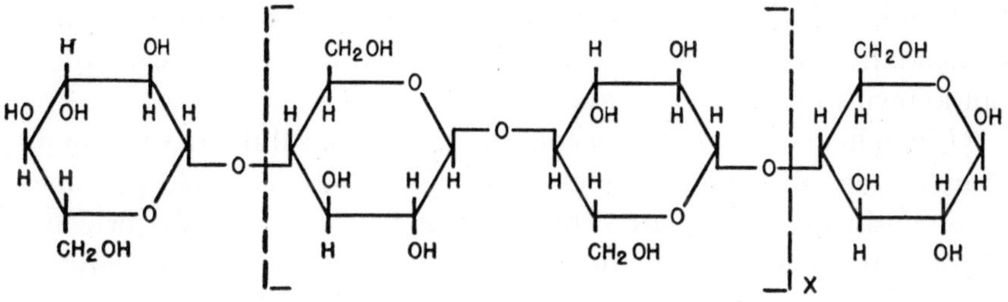

Figure 5–11. Cellulose formula.

the cellulose molecule in cotton consisting of 500 to 1000 such groups. The actual number in a particular case depends on such factors as growth conditions and degradation.

Cellulose is rather inert chemically and shows no pronounced acidic or basic reactions. Due to the preponderance of hydroxyl (—OH) groups in the molecule, the most common reaction into which cellulose enters is that of esterification, such as the well-known reactions with acetic acid and nitric acid to form cellulose acetates and cellulose nitrates.

In the absence of acidic or basic groups, pure cellulose would not be expected to enter into ion-exchange reactions such as, for example, the replacement of a hydrogen ion from the cellulose by a cation from an electrolyte solution. An extensive study of ion-exchange in cellulosic electrical-insulating papers by McLean and Wooten[49, 50] indicates considerable ion-exchange capacity in such materials, particularly for cations. However, they show that the cation-exchange capacity in such cases is due to acidic groups associated with non-cellulosic constituents in the papers. Cotton papers used in their studies probably contained considerably more non-cellulosic material than do processed cotton fabrics.

---

[49] McLean, D. A., and Wooten, L. A., *Ind. Eng. Chem.*, **31**, 1138–43 (1939).
[50] McLean, D. A., *Ind. Eng. Chem.*, **32**, 209–13 (1940).

Sookne and coworkers,[51, 52] working with dewaxed cotton and with depectinized cotton, found most of the cation-combining power of the cotton to be attributable to pectic noncellulosic material on the surfaces of the fibers. In samples of cotton dewaxed with alcohol and thoroughly washed with water, they found a cation-combining power of about 0.06 milliequivalent per gram of fiber. In depectinized cotton they found the cation-combining power to be of the order of 0.01 milliequivalent per gram of fiber. The effect of degradation, as by oxidative bleaching, on the ion-exchange capacity of cellulose is not known to have been studied.

Linen in general probably contains more noncellulosic constituents than cotton although, in high-grade bleached material, the linen may consist of nearly pure cellulose. The possibility for the presence of constituents which will enter into ion-exchange reactions is greater in linen than in cotton, although specific study of such a possibility is not known to have been made. The previously mentioned work of McLean and Wooten included linen-insulating papers, but the nature and extent of noncellulosic constituents in linen paper cannot be expected to be comparable with those of linen fabrics as encountered in laundering.

Wool fiber is classed as the protein, *keratin*, and is actually composed of several chemically distinct substances. Basically, the wool protein consists of the condensation products of a large number of amino acids, wherein the various amino acids form so-called polypeptide chains, and wherein adjacent polypeptide chains are held

$$-\mathrm{HN}-\underset{\mathrm{R}}{\underset{|}{\mathrm{CH}}}-\mathrm{CO}-\mathrm{HN}-\overset{\mathrm{R}}{\overset{|}{\mathrm{CH}}}-\mathrm{CO}-\mathrm{HN}-\underset{\mathrm{R}}{\underset{|}{\mathrm{CH}}}-\mathrm{CO}-\mathrm{HN}-\overset{\mathrm{R}}{\overset{|}{\mathrm{CH}}}-\mathrm{CO}-$$

together by cross-linkages. The true detailed structure of wool protein, as of all proteins, is still highly problematical and will not be discussed.

The characteristic of wool fiber that is particularly pertinent to a discussion of ion exchange is its amphoteric nature, that is, its combination of acidic and basic properties. Because of the presence of both acidic and basic groups in the keratin "molecule," wool fiber may react chemically with both acids and bases and may combine

[51] Sookne, A. M., and Harris, M., *J. Research Natl. Bur. Standards*, **25**, 47–60 (1940).

[52] Sookne, A. M., Fugitt, C. H., and Steinhardt, J., *J. Research Natl. Bur. Standards*, **25**, 61–9 (1940).

at its acid groups with cations or at its basic groups with anions. Thus, wool readily enters into typical ion-exchange reactions, with either cations or anions or with both simultaneously.

In studies of the combination of wool protein with hydrochloric acid and potassium hydroxide, Steinhardt and Harris[53] report the maximum acid-binding capacity of the wool, independent of ionic strength, to be 0.82 millimole per gram and the maximum base-binding capacity to be greater than 0.78 millimole per gram. In terms of general ion exchange, these results indicate the maximum cation-combining and anion-combining powers of wool to be of the order of 0.8 milliequivalent, each, per gram.

Sookne and coworkers[52] report the cationic ash content of wool to be very susceptible to the way in which the wool is handled. Thus, in the processing of wool, values of cationic ash as high as 0.4 milli-equivalent per gram of wool fiber were found immediately after soda ash treatment. In the case of wool fiber thoroughly washed and soured with acid, cationic ash values as low as 0.015 milliequivalent per gram were found. In this latter case, it is evident that H^+ ions from the sour have replaced the metallic cations previously held by the wool.

The place of ion exchange in a discussion of detergency and of the action of detergent solutions may now be considered. McBain[37] has indicated that ion exchange may be a real factor in removing soil that is in itself electrolytic or is derived from ionizable material. Anion-active substances, such as soaps and synthetic anionic detergents, and cation-active substances are capable of entering into ion-exchange reactions in the same way as simple electrolytes.

It is foreseeable, although not actually studied heretofore, that soil which is held to fibers by actual chemical combination at active groups in the fibers should be subject to replacement by the mechanics of ion exchange under suitable conditions. Thus, the anionic fatty group resulting from the dissociation of soap in water might well be capable of replacing anionic soil chemically combined on fabric at a basic group in the fiber structure. On the other hand, the electrolytic cation from soap, such as Na^+, may have a similar action with cationic soil, particularly when such cations are fortified by like ions from a builder.

---

[53] Steinhardt, J., and Harris, M., *J. Research Natl. Bur. Standards*, **24**, 335–67 (1940).

Substitution of detergent ions or ionic groups for soil ions or ionic groups at chemically active centers in fibers may also result in new combinations in the detergent solution proper. Thus, as early as 1909, Spring[54-56] postulated soil attached to fabric as a pseudo-chemical sorption compound which, in the presence of soap solution, decomposes to yield a soil-soap sorption compound which passes into the detergent solution. McBain[37] suggests an alternative involving double decomposition in which two sorption compounds are formed, as represented schematically by the equation

$$\text{fabric} \cdot \text{soil} + \text{soap} \rightleftharpoons \text{fabric} \cdot \text{soap} + \text{soil} \cdot \text{soap} \tag{8}$$

It is further conceivable that the (soil·soap) compound thus formed might in some cases be stoichiometric rather than sorptive, and that a consequent stabilization of the soil in the detergent can thus be realized to reduce the reversal of the reaction of the equation. As a hypothetical example of such a case, a cationic soil group, replaced on the fabric by the Na^+ ion from a soap solution, might react with the anionic fatty group of the soap to form a new undissociated or only slightly dissociated combination.

Another manifestation of ion exchange closely related to detergency involves the removal of dye from fabric (fading of colored work). In this process, which has much significance in laundry washroom operations, dyes chemically combined with fabric as either cationic or anionic groups may, under some conditions, be undesirably replaced by cations or anions from the detergent solution. The removal of dyes by ion exchange is sometimes further abetted by solubilization of the dye within detergent micelles. Basic dyes are more affected by anion-active detergents than are acid dyes.

Stabilization

Stabilization, as an action of detergent solutions, pertains to the maintenance of solid and liquid soil in a dispersed condition in the deterging medium. The stabilization of emulsions of oil has already been discussed to a considerable extent in connection with the action of emulsification. Discussion of stabilization of dispersed solid soil has already been introduced as a topic somewhat analogous to peptization.

---

[54] Spring, W., *Rec. trav. chim.*, **28,** 120–35 (1909).
[55] Spring, W., *Z. Chem. Ind. Kolloide,* **4,** 161–8 (1909).
[56] Spring, W., *Bull. Acad. roy. Belg.*, **1909,** 187–206.

The system of liquid and solid soil dispersed in a detergent solution, as obtained in laundry washing operations, is typically a colloidal system, even though some of the dispersed droplets or particles only approach true colloidal size.

As will be evident from Chapter 10, those liquid and solid soils which most necessitate the use of soap or other detergent are, in general, hydrophobic materials; that is, their "attraction" to pure water is only slight at best. Such soils would be the neutral oils and most earthy material, dispersions of which in water would be expected to show little solvation. On the other hand, fatty acid soils, particularly when neutralized by the detergent, and proteinaceous soils are hydrophilic; that is, although they do not enter into true solution in water, there exists between them and the water an "attraction" which permits them in a dispersed state to hold an incasement of water by solvation.

A dispersion of solid soil in pure water can be considered to consist of particles usually negatively charged because of the high dielectric constant of the water (see Chapter 7). Some (the hydrophilic particles) are incased in water bound by solvation, and some (the hydrophobic particles) are solvated very little. A dispersion of solid soil in anionic detergent solution, in the absence of builders, will likewise consist of negatively charged droplets and particles. However, in this latter case the droplets and particles will be incased in an adsorbed film of oriented detergent. Those which, in water alone, were hydrophobic will now be hydrophilic because of the adsorbed detergent, with the result that the solvation of all the droplets and particles will be pronounced.

The stabilization of dispersed material by an electric charge has already been mentioned in connection with emulsification and peptization. A dispersed system wherein all the dispersed particles or droplets carry like charges will tend to be stabilized by the electrical repulsion between the particles or droplets. Neutralization of these charges toward the isoelectric point, as by the addition of an electrolyte, reduces the stability of the dispersion. However, such an addition of electrolyte as will carry the system quickly past the isoelectric point and reverse the charge on the droplets or particles will reestablish stability in the system.

In a detergent solution as the dispersion medium, the droplets and particles are not only charged but also they are solvated by reason of the presence of the detergent. The charge and the solvation, separately or together, aid stabilization. Thus, it has been shown in a system somewhat analogous to that obtaining in a washwheel, that the flocculation of gelatin by the addition of tannin is not a case of electrical neutralization. De Jong[57,58] shows that the flocculation in this case is due to removal of water of solvation from the gelatin by the tannin, whereby electrolytes already present may effect the flocculation. In other words, when the gelatin is solvated, the solvation furnishes protection against any flocculating action by electrolytes that are also present. A similar condition has been noted by the writer in unpublished work, in the case of fairly stable suspensions of clay in water whereby the addition of tannin, itself a neutral colloid, caused very pronounced and rapid flocculation of the clay. This can be considered as due to de-solvation of the clay particles by the tannin and then flocculation by the alkali salts associated with the clay and thus present in solution.

Probably the most significant contribution of soaps and other detergents to the stability of dispersed systems is that already mentioned, namely, the furnishing of a hydrophilic protective "coating" to the dispersed phase such as will enhance solvation. However, this very likely is not the only part they play in stabilization. Bearing in mind that all except nonionic detergents are pronounced electrolytes in aqueous solution, it may well be that ionic detergents impart electrolytic effects on dispersed systems in a manner analogous to that of electrolytic builders as will be discussed in greater detail in Chapter 6. We know that maintenance of specific alkalinity in a dispersion of negatively charged particles favors stabilization of the particles. This is no doubt associated with powerful adsorption of hydroxyl ion, in comparison to adsorption of many metallic cations. If we consider that, from a solution of appreciable alkalinity wherein hydroxyl ions are readily available, hydroxyl ions may tend to be adsorbed to the exclusion of cations, we may see that the flocculating action of the cations might be counteracted to an appreciable extent. It has been determined experimentally, for example, that potassium hydroxide is much less of a flocculating agent for negatively charged colloid than is potassium chloride, even though the same flocculating

---

[57] De Jong, H. G. B., *Chem. Weekblad*, **20**, 260 (1923).
[58] De Jong, H. G. B., *Rec. trav. chim.*, **42**, 437–72 (1923).

agent (K^+ ion) is present in both cases. Further, many alkaline electrolyte salts show specific stabilizing action for negatively charged dispersions. It appears to be only a short step to extend this reasoning at least to soaps, the alkalinity of whose solutions might well exert a comparable stabilizing effect. Here lies a definite gap in our confirmed knowledge of detergent action. Filling the gap may provide valuable information as to why some agents of comparable surface activity differ widely with respect to detergent power.

6

Nature and Properties of Builder Solutions

Nature of Builders

The so-called builders used in laundry washing as adjuncts to soap and synthetic detergents may be divided into several distinct chemical classes, as follows:

(a) Alkali salts, usually the sodium salt, of weak inorganic acids;
(b) Sodium or potassium hydroxide;
(c) Neutral inorganic salts, usually the sodium salt, of strong inorganic acids.

Various proprietary combinations of these compounds are marketed as "tailored" builders. In addition, there are other materials which, upon addition to detergent solutions, produce effects similar to those of the usual builders. These materials include sodium carboxymethyl cellulose, fatty acids, and even detergents of one composition when added to a solution of a detergent of another composition. However, discussion of these latter materials will be reserved for Chapters 8 and 9.

The alkali salts of weak inorganic acids most commonly used in laundering are:

(a) Sodium carbonate (soda ash), Na_2CO_3;
(b) Modified soda (sodium carbonate plus sodium bicarbonate), $Na_2CO_3 + NaHCO_3$;
(c) Sodium sesquicarbonate, $Na_2CO_3 \cdot NaHCO_3 \cdot 2H_2O$;
(d) The sodium silicates;

(e) The sodium phosphates;
(f) Sodium tetraborate (borax), $Na_2B_4O_7 \cdot 10H_2O$.

The potassium salts are seldom used because of considerably greater cost.

The alkali hydroxides (almost exclusively sodium hydroxide) are sources of high alkalinity and high unbuffered pH and must be used with care as builders because of possible deleterious action on many fabrics.

The neutral inorganic salts most commonly used as builders are sodium chloride, NaCl, and sodium sulfate, Na_2SO_4. These salts dissociate but do not hydrolyze, and yield practically neutral aqueous solutions. They are used particularly in connection with anionic synthetic detergents. Although the neutral alkaline earth salts, such as calcium and magnesium chloride and magnesium sulfate, have many properties in common with the corresponding sodium salts, they are seldom, if ever, used as builders. Their use with soaps would be objectionable because of the formation of insoluble alkaline earth soaps. However, in connection with synthetic detergents which are not precipitated by calcium or magnesium ion, their use can be beneficial and natural "hardness" in wash water can serve as a desirable builder in some cases.

Hydrolysis of Builders

Hydrolysis of the sodium salts of weak inorganic acids is undoubtedly the most significant *chemical* reaction into which they enter in so far as laundering and detergency in general are concerned, and will be discussed for each case in detail.

Sodium carbonate (soda ash) reacts in aqueous solution in the following manner:

$$Na_2CO_3 \rightleftharpoons 2Na^+ + CO_3^= \quad (1)$$

$$CO_3^= + H_2O \rightleftharpoons OH^- + HCO_3^- \quad (2)$$

$$HCO_3^- + H_2O \rightleftharpoons OH^- + H_2CO_3 \quad (3)$$

The OH^- of equation (2) accounts for the increase in hydroxyl-ion concentration found upon dissolving sodium carbonate in water. Shields[1] and Kölichen[2] have measured the extent of hydrolysis of

[1] Shields, J., *Phil. Mag.* (5), **35**, 365 (1893); *Z. physik. Chem.*, **12**, 174 (1893).
[2] Kölichen, K., *Z. physik. Chem.*, **33**, 173 (1900).

sodium carbonate at different concentrations with the results shown in Table 6–1. As is to be expected from the mass action effect, the

TABLE 6–1. HYDROLYSIS OF SODIUM CARBONATE IN WATER [3]

Gram-Mol. Na_2CO_3	0.942[1]	0.19[1]	0.094[2]	0.0477[2]	0.0238[2]
Per cent hydrolysis	0.53	2.12	3.17	4.87	7.10

extent of hydrolysis increases with decreased concentration. The degree of hydrolysis at a given concentration increases with increased temperature.

Modified soda consists of a physical mixture of sodium carbonate and sodium bicarbonate, in approximately equimolar proportion. Bicarbonate ion (HCO_3^-) from the sodium bicarbonate will react with water as shown by equation (3). In the presence of sodium carbonate, however, sodium bicarbonate acts as would an acid and tends to reverse the reaction illustrated by equations (1) and (2). The result is then lesser hydrolysis and lower hydroxyl-ion concentrations in solutions of modified soda than in solutions of sodium carbonate.

Sodium sesquicarbonate is a double salt of sodium carbonate and sodium bicarbonate, probably in equimolar proportions. It separates readily in water solution into the carbonate and bicarbonate and therefore may be expected to show much the same reactions as does modified soda. It is questionable whether the sesquicarbonate referred to frequently in literature on detergency is always the true double salt or rather is simply modified soda.

The *sodium silicates* may be varied in composition more or less at will, according to the proportion of, say, sodium carbonate to silicon dioxide used as starting materials for subsequent production of the silicate by fusion. The most basic of this group of builders is sodium metasilicate, Na_2SiO_3, probably written more properly as $Na_2O \cdot SiO_2$. By increasing the proportion of silica to sodium oxide, silicates up to about $Na_2O \cdot 4SiO_2$ have been studied and used as builders. Silicates wherein the ratio of silica to sodium oxide ranges from 2 to 4 constitute the well-known "water glass."

In the presence of water the glassy silicates are hydrolyzed as illustrated by the following reaction for sodium metasilicate:

$$Na_2O \cdot SiO_2 + H_2O \rightleftharpoons Na^+ + OH^- + H_2SiO_3 \quad (4)$$

[3] Mellor, J. W., "A Comprehensive Treatise on Inorganic and Theoretical Chemistry," Vol. II, p. 762, New York, Longmans, Green and Co., 1937.

The reversal of this reaction is demonstrated by the reaction between sodium hydroxide and silicic acid or even silica to form a sodium silicate in the presence of water.

The disposition of the silicic acids resulting from the hydrolysis of sodium silicates is of considerable consequence. These acids, of the general formula $m\mathrm{SiO_2} \cdot n\mathrm{H_2O}$, form colloidal micelles in water, extending even to gels at sufficiently high concentration. These micelles are of uncertain composition and are highly solvated. Their significance in detergency will be discussed in Chapter 9.

The *sodium phosphates* of most importance as builders are trisodium orthophosphate, tetrasodium pyrophosphate, and sodium hexametaphosphate, represented by the formulas Na_3PO_4, $Na_4P_2O_7$, and $Na_6P_6O_{18}$, respectively. The composition of at least the latter of these phosphates is very complex because of the possibilities for isomerism within the molecule.

Hydrolysis of the first two of the above-listed sodium phosphates in water may be illustrated as follows:

Trisodium orthophosphate—

$$\mathrm{Na_3PO_4 + H_2O \rightleftharpoons Na_2HPO_4 + Na^+ + OH^-} \quad (5)$$

$$\mathrm{Na_2HPO_4 + H_2O \rightleftharpoons NaH_2PO_4 + Na^+ + OH^-} \quad (6)$$

$$\mathrm{NaH_2PO_4 + H_2O \rightleftharpoons H_3PO_4 + Na^+ + OH^-} \quad (7)$$

Tetrasodium pyrophosphate[4]—

$$\mathrm{Na_4P_2O_7 + H_2O \rightleftharpoons 2Na_2HPO_4} \quad (8)$$

The disodium acid orthophosphate of equation (8) further hydrolyzes in accordance with equations (6) and (7).

Bell[4] considers that sodium hexametaphosphate in solution at 100°C is partly hydrolyzed to orthophosphate and partly depolymerized to trimetaphosphate, according to

$$\mathrm{3(NaPO_3)_6 + 12H_2O \rightleftharpoons 2(NaPO_3)_3 + 12NaH_2PO_4} \quad (9)$$

Trimetaphosphate then would react with water, as

$$\mathrm{(NaPO_3)_3 + H_2O \rightleftharpoons Na_3H_2P_3O_{10}} \quad (10)$$

$$\mathrm{Na_3H_2P_3O_{10} + 2H_2O \rightleftharpoons 3NaH_2PO_4} \quad (11)$$

The sodium dihydrogen orthophosphate of equations (9) and (11) may hydrolyze to yield hydroxyl ion according to equation (7).

[4] Bell, R. N., *Ind. Eng. Chem.*, **39**, 136–40 (1947).

However, the simple composition for sodium hexametaphosphate shown in equation (9) as $(NaPO_3)_6$ is somewhat questionable. The probable existence of isomers of this phosphate, such as $Na_5(NaP_6O_{18})$, $Na_4(Na_2P_6O_{18})$, and $Na_2(Na_4P_6O_{18})$, is indicated by such properties as its ability to sequester alkaline earth cations.

Sodium tetraborate (borax), $Na_2B_4O_7 \cdot 10H_2O$, or more properly written $Na_2O \cdot (B_2O_3)_2 \cdot 10H_2O$, is only sparingly soluble in water at ordinary temperatures but is readily soluble at 62°C where conversion to the pentahydrate takes place. Hydrolysis of solutions of the salt is relatively slight, a $0.1N$ solution of the sale containing only about $2 \times 10^{-5}N$ hydroxyl ion, corresponding to about 0.3 per cent hydrolysis. Its hydrolysis may be illustrated as follows:

$$Na_2B_4O_7 + 3H_2O \rightleftharpoons 2Na^+ + 4HBO_2 + 2OH^- \quad (12)$$

Before leaving the discussion of hydrolysis of alkaline builders, two closely related properties, neutralizing value and availability of alkalinity, will be introduced. These properties pertain both to the alkali salts of weak acids and to the alkali hydroxides. The significance of these properties in detergency was advanced originally by Snell,[5, 6] who considers that the usefulness of an alkaline builder in the wash-wheel continues only so long as it maintains an alkalinity in the solution greater than that corresponding to the alkalinity of the soap or other detergent which it is building. Snell reports the pH values and hydroxyl-ion concentrations shown in Table 6–2 for 0.033 per cent

TABLE 6–2. INITIAL ALKALINITY IN SOLUTIONS OF BUILDERS [5]
(at 25°C)

	0.033% *Solution of Builder*	
	pH	$C_{OH^-} \times 10^3$
Sodium hydroxide	11.85	7.10
Sodium metasilicate, $Na_2O \cdot SiO_2 \cdot 9H_2O$	11.2	1.60
Trisodium phosphate, $Na_3PO_4 \cdot 12H_2O$	10.8	0.63
Sodium silicate, 1 : 1.58 (anhydrous)	10.7	.50
Sodium carbonate, Na_2CO_3	10.65	.45
Sodium silicate, 1 : 3.86 (anhydrous)	10.1	.12
Modified soda, $Na_2CO_3 + NaHCO_3$	10.0	.10
Borax, $Na_2B_4O_7 \cdot 10H_2O$	9.35	.022
Sodium oleate (soap)	10.2	.16

solutions of various builders at 25°C, this concentration being representative of those used in laundering. The electrometric titration

[5] Snell, F. D., *Ind. Eng. Chem.*, **24,** 76–80 (1932).
[6] Snell, F. D., *Ind. Eng. Chem.*, **25,** 1240–6 (1933).

curves for 0.66 per cent solutions shown in Figure 6–1 are also taken from Snell's work, the higher concentration being necessary to assure sufficient accuracy with the hydrogen electrode used for the measurements. Snell defines neutralizing value as a measure of the extent to which a builder will neutralize acidity in the dirt before it becomes ineffective for that purpose, and considers that the value should be measured above the neutral point for the particular soap (pH 10.2 for 0.033 per cent sodium oleate). Neutralizing values for 0.66 per cent builder solutions can be taken directly from the curves on Figure 6–1,

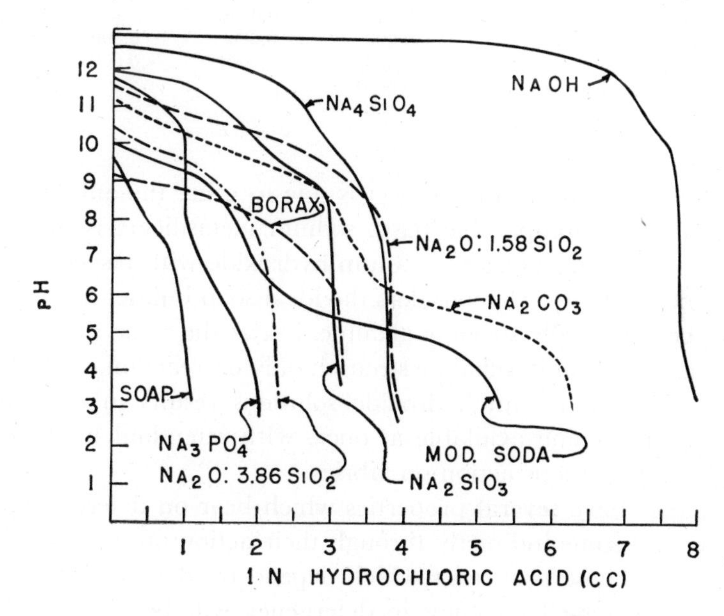

Figure 6–1. Electrometric titrations (expressed as pH) of 0.66% solutions of common builders in comparison to 0.66% sodium oleate.[5]

and can be obtained for other concentrations by preparing similar electrometric titration curves for the particular concentrations.

Solutions of the alkali salts of weak acids are buffered. The concentration of free alkali from hydrolysis when the salt is first dissolved does not represent the total alkali available for reaction with acid, since further hydrolysis occurs and additional alkali is formed as reaction with acid removes the previously formed alkali. Alkalinity available from the builder below the pH value for the particular neutral soap used is of no value in connection with building of that soap because of decomposition of the soap at pH values below its neutral point. Snell reports a comparison of 0.033 per cent solutions of different builders after reaction with varying amounts of acid, as shown in Table 6–3. He points out that the apparent superiority of

TABLE 6–3. CONCENTRATION OF HYDROXYL ION [5]

(After reaction with varying amounts of acidic materials from fabrics)

	Moles acidic material removed			
	0.0005	0.001	0.0015	0.002
Sodium metasilicate	0.008	0.001	0.0043	<0.0001
Trisodium phosphate	0.003	<0.0001		
1 : 1.58 sodium silicate	0.002	0.006	0.0002	0.0001
Sodium carbonate	0.0006	0.0001	<0.0001	
1 : 3.86 sodium silicate	<0.0001			
Modified soda	<0.0001			
Borax	<0.0001			

the 1 : 1.58 silicate in this respect is due to using the anhydrous salt and that, on a comparable basis, sodium metasilicate is superior to all the other builders except sodium hydroxide with respect to availability of alkalinity. He considers the ideal soap builder as one which would be potentially strongly alkaline, but at the same time buffered so as to yield all of its alkaline strength only on reaction. The absence of buffering in sodium hydroxide solutions results in most of the alkalinity becoming available at once, with attendant high pH and danger of deleterious action on fabrics.

Builders have several properties which bear on detergency, some directly and some indirectly through their action on soaps and other detergents. Those properties which depend on their action on soaps and the like for effectiveness in detergency will be discussed in the next two chapters. The properties which make builders detergents in their own right are discussed below.

Neutralization by Builders

Those builders which produce alkaline solutions may neutralize acid soil, including conversion of high-molecular-weight fatty acid soil to soap, in the following manner:

$$\underset{\text{(fatty acid soil)}}{RCOOH} + Na^+ + OH^- \rightleftharpoons \underset{\text{(soap)}}{RCOONa} + H_2O \qquad (14)$$

This type of reaction is significant in detergency in several respects: it converts the fatty acid soil to a more soluble material—soap; it supplies new soap within the interstices of the fiber to exert detergent action on other soil; it is the basis of the action of spontaneous emulsification of oily soil as will be discussed below.

The significance of saponification, the cleavage of fatty esters by alkalies, under conditions of laundering is questionable. It is true that the extent to which fatty soil can be saponified by alkalies at the temperatures, concentrations, and time obtaining in the washwheel is very slight. On the other hand, assuming that even a very slight amount of soap is produced under these conditions, that soap will be disposed where it can be most effective for detergency, namely, within the soil that is to be deterged. It is not known that any actual study has been made of the possible significance of even slight amounts of soap produced under such conditions.

Surface Activity of Builder Solutions

Numerous studies have been made of builders as surface-active substances, from the standpoint of surface- and interfacial-tension depressants, wetting agents, emulsifiers, and foaming agents. In general, the builders do not appreciably affect the surface tension of water against air and it is doubtful that they exert any significant influence on the interfacial tension between water and inert neutral oils. Thus, Richardson[7] found that the drop number of solutions of $Na_2O \cdot 2.83SiO_2$, in concentrations of 0.05 to 0.25 per cent, against kerosene was not different from that for water alone. On the other hand, the interfacial tension between water and oily soils containing neutralizable or saponifiable material can be materially reduced by the presence of a builder, through the formation of soap by reaction between the reactive component of the soil and the hydrolysis alkalinity of the builder. With reference to the wetting power of builder solutions in the absence of soap, Baker[8] has reported the data shown in Figure 6–2 for various builder solutions against glass. In his work, 0.31 cc portions of the builder solution were placed on glass at 20°C and the radii of the zones of spread were then measured. Although his data indicate a superiority of sodium metasilicate in the particular application, the parallelism between wetting of siliceous glass and wetting of textile fibers is not known. Baker has also studied the dis-

[7] Richardson, A. S., *Ind. Eng. Chem.*, **15**, 241–3 (1923).
[8] Baker, C. L., *Ind. Eng. Chem.*, **23**, 1025–32 (1931).

placement of oil films from glass surfaces by builder solutions. It is probable that the wetting action of builder solutions in such cases is more pertinent to industrial and general household cleansing than to laundering. Powney and Frost[9] report that the contact angle made by water against paraffin wax and air is unaffected by the addition of caustic soda to the water.

Shorter[10] pointed out in 1916 that the soap resulting from neutralization of fatty acids in the soil by alkali in the detergent solution is very effective as a detergent for neutral oil because "it is on the spot where it is wanted." Harkins and Zollman[11] conducted a more

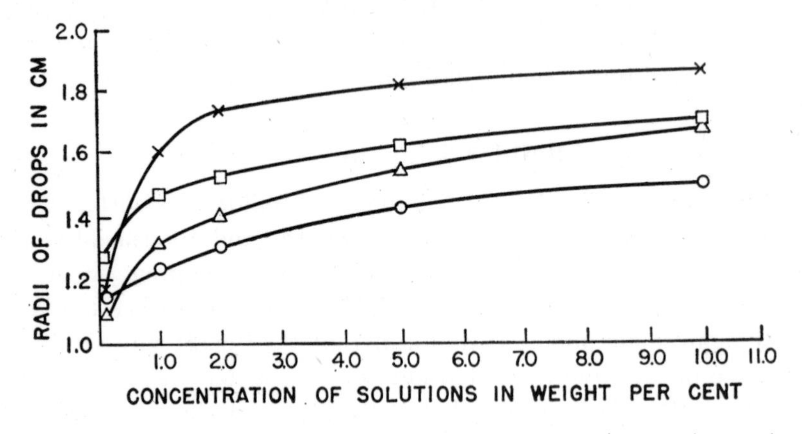

Figure 6–2. Radii of areas of coverage of 0.31 cc. drops of builder solutions on a horizontal glass plate.[8]

×—$Na_2SiO_3 \cdot 5H_2O$
△—$Na_3PO_4 \cdot 12H_2O$
□—NaOH
○—Na_2CO_3

specific study of this factor, using sodium hydroxide in water against oleic acid in benzene. The interfacial tensions in these cases were found to be very small: 0.38 dyne for 0.005*N*, 0.31 dyne for 0.01*N*, and 0.6 dyne for 0.1*N* solutions in both phases. These values are much lower than those obtained where the starting solute is sodium oleate in concentrations even considerably greater than that formed in the present case by reaction between the sodium hydroxide and the oleic acid. Harkins and Zollman consider this to suggest that the energy of

---

[9] Powney, J., and Frost, H. F., *J. Textile Inst.*, **28,** T237–54 (1937).
[10] Shorter, S. A., *J. Soc. Dyers Colourists*, **32,** 99–108 (1916).
[11] Harkins, W. D., and Zollman, H., *J. Am. Chem. Soc.*, **48,** 69–80 (1926).

chemical union between the base and the acid may be used directly to aid in the formation of the surface; in other words, less outside energy need be employed to effect surface formation.

Emulsification by Builders

Kling and Schwerdtner[12] observed under a microscope the action of aqueous solutions of several surface-active agents on droplets of oleic acid. In some cases, spontaneous emulsification of the oleic acid appeared as a smooth "corrosive" action, progressing inward, and in other cases the droplets underwent spontaneous violent action and appeared to "shatter." Harkins and Zollman[11] have observed, in the case of water and benzene, that when the interfacial tension is below 10 dynes per cm. the benzene emulsifies easily in the aqueous phase, and that below 1 dyne it appears to emulsify spontaneously. Although this process of spontaneous emulsification has been studied very little in connection with detergency, it is probable that the mechanics of the action involves the neutralization of the fatty acid by alkali present from hydrolysis of the surface-active agent.

The foregoing discussion leads to a probable role of alkaline builders in detergency, involving the neutralization of acid soil. Assume a soil of neutral oily material, interspersed with fatty acid, in contact with a solution of alkaline builder. The sodium and hydroxyl ions present in the solution by virtue of hydrolysis of the builder can be expected to react with the free fatty acid in the soil to form soap which will then be interspersed through and in intimate contact with the neutral soil. The result is a pronounced reduction in interfacial tension between the neutral oil and the builder solution, which can conceivably bring about spontaneous emulsification of the oil under some conditions. In this connection, the builders or their hydrolysis products are not in themselves particularly effective emulsifying agents for neutral oils. The presence of fatty acids or saponifiable material and the use of builders which yield alkaline solutions are both essential to the pronounced functioning of builders as emulsifiers. In such cases, the soaps formed by the reaction between the builder and the fatty acids or the saponifiable material are actually the real emulsifiers, rather than the builder proper.

---

[12] Kling, W., and Schwerdtner, H., *Melliand Textilber.*, **22**, 21–8 (1941).

The emulsifying powers of builders have been studied by Stericker,[13] Baker,[8] Vincent,[14] and Hillyer,[15] and others. In general the builders, especially the silicates of the more siliceous types, have some slight emulsifying powers in their own right, but this power is very weak in comparison to that of soaps and other detergents.

Foaming of Builder Solutions

The more commonly used builders, with the exception of the silicates, cannot be considered as foaming agents in the concentrations used in laundering. Previously unpublished results of studies[16] of the foaming of aqueous solutions of several electrolyte builders other than silicates are shown in Table 6–4. It may be seen that in some

TABLE 6–4. FOAMING OF AQUEOUS BUILDER SOLUTIONS [16]

Builder	Foam time, sec.[a]	
	0.2% solution	1.0% solution
Sodium bicarbonate	Nonfoaming	Nonfoaming
Sodium carbonate	4	4
Trisodium phosphate	4	6
Tetrasodium pyrophosphate	Nonfoaming	3
Sodium hexametaphosphate	Nonfoaming	Nonfoaming
Borax	Nonfoaming	4
Sodium sulfate	Nonfoaming	Nonfoaming
Sodium chloride	Nonfoaming	Nonfoaming
Sodium hydroxide	Nonfoaming	4

[a] 100 ml. of solution vigorously shaken in a stoppered flask.

Note: Under these conditions, distilled water alone had an apparent foam time of 1 to 2 seconds. "Nonfoaming" indicates no appreciable difference from distilled water.

cases there is a greater tendency for foaming at the higher concentration, but that in no case is either the extent of foaming or the persistency of the foam appreciable. The apparent slight superiority of trisodium phosphate is not considered to be significant.

In a study of the foaming power of sodium silicate solutions, Stericker[13] found that 0.37 per cent $Na_2O \cdot 3.3SiO_2$ produced a foam in distilled water at room temperature but that, at 65° to 70°C, 1.15 per cent was required. The foam secured with silicates of other ratios was also less stable at higher temperatures. At room temperature, the $Na_2O \cdot 3.3SiO_2$ gave a foam at a lower concentration than did other

[13] Stericker, W., *Ind. Eng. Chem.*, **15,** 244–8 (1923).
[14] Vincent, G. P., *J. Phys. Chem.*, **31,** 1281–315 (1927).
[15] Hillyer, H. W., *J. Am. Chem. Soc.*, **25,** 511–24 (1903).
[16] Niven, W. W., Jr., and Hathaway, J., unpublished data.

more alkaline silicates but, at 65°C, silicates with higher ratios of alkali were better. These findings of Stericker's are in line with the colloidal nature of silicate solutions and their consequent ability to produce films of heterogeneous composition, as previously discussed in connection with the foaming of soap solutions.

In the presence of oils, Stericker found that less of the silicates were required to produce foams. With a little spindle oil to which a small amount of fatty acid had been added, he found that the more alkaline the silicate (the higher the Na_2O/SiO_2 ratio) the less was required to produce a stable foam with the oil, using 2.35 per cent of the oil. Following are the optimum concentrations of various silicates found by Stericker to be required for persistent foam, as compared to sodium carbonate:

$Na_2O \cdot 2.4SiO_2$ — 0.6 — 0.8%
$Na_2O \cdot 2.1SiO_2$ — 0.4%
$Na_2O \cdot 1.75SiO_2$ — 0.2%
Na_2CO_3 — 0.5%

Aqueous solutions of sodium carboxymethyl cellulose, recently introduced as a builder for synthetic detergents,[17] show some tendency to foam when the concentration is sufficiently high, and the foam produced is quite persistent.[16] However, foaming in concentrations below 1.0 per cent is not extensive, nor is the persistency of the foam high.

Peptization by Builders

As has been pointed out by Gortner,[18] a peptizer "either (a) must have one ion in common with the material to be dispersed, or (b) must be capable of forming a soluble compound with the material to be dispersed, or (c) must have one ion which is very strongly adsorbed by the material being dispersed." Several of the common builders may fulfill one or more of these requirements. Peptization by these materials in the specific case of laundering is not known to have been studied. However, such actions as the well-known peptization of clays by alkalies and the peptization of gels by electrolytes in general should

[17] Vaughn, T. H., and Smith, C. E., *J. Am. Oil Chemists' Soc.*, **25**, 44–51 (1948).
[18] Gortner, R. A., "Outlines of Biochemistry," p. 28, New York, John Wiley & Sons, Inc., 1929.

have direct bearing on washroom operations. Any tendency for the adsorption of a builder ion on or in an aggregation of soil whereby the "solution pressure" of the aggregated units is increased as discussed in Chapter 5, will favor dispersion of the aggregation by the action of peptization.

Stabilization of Dispersed Systems by Builders

Probably the most significant detergent property of builders in the absence of soap or synthetic detergent is that of effecting the stabilization or deflocculation of dispersed systems. The terms **deflocculation** and **peptization** have been used interchangeably in many past references on builder solutions and have been considered to be synonymous.

Fall[19] has studied the deflocculating or soil-suspending powers of several silicates, trisodium phosphate, sodium hydroxide, and sodium carbonate, using suspensions of manganese dioxide. At a builder concentration of 0.025 per cent, their suspending powers descend in the order given, ranging from 445 cg. of MnO_2 per liter for $Na_2O \cdot 3.97SiO_2$ to 110 cg. for Na_2CO_3. A concentration of 0.15 per cent yielded the following values in centigrams of MnO_2 per liter at 40° and 75°C, respectively: $Na_2O \cdot 3.97SiO_2$, 396, 287; $Na_3PO_4 \cdot 12H_2O$, 70, 274; NaOH, 17, 52; Na_2CO_3, 0.000, 12. Fall also studied the relative suspending powers of three silicates with SiO_2/Na_2O ratios of 1.62, 2.82, and 3.97, and found them all to be equally effective at optimum concentration.

Baker[8] somewhat paralleled Fall's work to include sodium metasilicate, and determined the deflocculating powers of builder solutions from the standpoints of both their concentrations and their pH values. Using purified bone black as the dispersed phase, he found the dilute solutions to be more effective deflocculators than the concentrated solutions. Based on the maximum concentration at which deflocculation is still effective, the builders studied may be arranged in the order $Na_2O \cdot SiO_2 \cdot 5H_2O \cong Na_3PO_4 \cdot 12H_2O > Na_2O \cdot 3.25SiO_2 > Na_2CO_3 > NaOH$. On the basis of maximum pH value, the order is $Na_2O \cdot SiO_2 \cdot 5H_2O > NaOH \cong Na_3PO_4 \cdot 12H_2O > Na_2CO_3 > Na_3O \cdot 3.25SiO_2$. The sodium metasilicate was the only builder to show appreciable stabilization of bone black above a pH of 11.0.

---

[19] Fall, P. H., *J. Phys. Chem.* **31,** 801–49 (1927).

Prevention of redeposition of soil by various builders was studied by Carter[20] as a measure of the deflocculating power of the builders. Using ferric oxide and other pigments as the suspended soil, he found dilute solutions to greatly decrease the deposition of the soil from the suspension onto fabric. He further found NaOH, Na_2CO_3, and modified soda to have only slight power, if any, to prevent deposition of such materials, and $Na_3PO_4 \cdot 12H_2O$ to have a power intermediate between that of the silicates and that of the others. The matter of redeposition of soil onto fabrics and its prevention will be considered in detail in Chapter 13.

Snell[21] also studied the deflocculating powers of various builders without added soap, using oil-coated burnt umber as the dispersed phase, but the umber contained free fatty acid in the oil. His results, as shown in Table 6–5, therefore reflect to some extent the effect of soap produced by interaction between the alkali and the free fatty acid.

TABLE 6–5. UMBER SUSPENDED AFTER 24 HOURS[21]

(From shaking 1 gram of oiled umber at 20°C with 100 cc of the builder solution)

Builder (0.1% solution)	Umber/100 cc suspension, mg.
Sodium hydroxide, NaOH	18.9
Sodium metasilicate, $Na_2O \cdot SiO_2 \cdot 9H_2O$	32.2
Trisodium phosphate, $Na_3PO_4 \cdot 12H_2O$	25.3
Sodium carbonate, Na_2CO_3	12.2
Modified soda, $Na_2CO_3 + NaHCO_3$	12.2
Sodium oleate soap (for comparison)	48.6

In an extensive study of the deflocculation of suspensions of earthy soil in the presence of naturally occuring electrolytes, the Corps of Engineers, U. S. Army[22] have reported the results shown in Table 6–6 for suspensions of 45.2 g. of clayey soil in 1130 cc of suspension. These results show very pronounced deflocculating powers for sodium metasilicate, trisodium phosphate, and sodium bicarbonate. Although the conditions were much different from those encountered in laundering because of the high concentration of the clay suspensions, the results serve to illustrate the deflocculating power of certain alkali salts for clayey material. Such material, though not studied by others

[20] Carter, J. D., *Ind. Eng. Chem.*, **23**, 1389–95 (1931).

[21] Snell, F. D., *Ind. Eng. Chem.*, **25**, 162–5 (1933).

[22] U. S. Army, Corps of Engineers, "Notes on Principles and Applications of Soil Mechanics," Appendix II, Fort Peck, Montana (1939).

in connection with builders for detergency, may be a very significant soil with which to contend in laundering.

The contrast between the results reported above in connection with clayey soil and those discussed in connection with suspensions of such materials as MnO_2, Fe_2O_3, burnt umber, and bone black, points to a possibility that the nature of the dispersed material may have considerable bearing on the extent of deflocculation realized with various agents. It is logical to expect that the relative "hydrophilicity" of different suspended materials should dictate to a considerable extent the ease with which they can be deflocculated.

TABLE 6–6. EFFECT OF ADDITION AGENTS ON STABILITY OF CLAY SUSPENSIONS [22]

Agent	Amount of agent added, g./100 cc	Stability
None		Quite rapid floc
NaOH	0.008	Immediate dense floc
Na_2SiO_3	.027	Light floc at 1 hour
	.054	Light floc at 15 minutes
	.081	Light floc at 15 minutes
	.108	Light floc at 30 minutes
Na_2CO_3	.040	Immediate dense floc
	.080	Immediate dense floc
	.120	Immediate dense floc
Na_3PO_4	.010	Light floc at 1 hour
	.020	Light floc at 24 hours
	.040	Light floc at 30 minutes
	.080	Light floc at 15 minutes
$NaHCO_3$	.020	Light floc at 24 hours
	.040	Light floc at 2 hours
	.080	Light floc at 15 minutes

As has been pointed out by Powney and Wood,[23] a factor of great importance in the stabilization of emulsions or suspensions of solids is the magnitude of the zeta potential of the droplets or particles. With this in mind they studied the effect of various alkalies on the electrophoretic mobility of oil drops dispersed in water. Their data with highly refined mineral oil are shown in Figure 6–3. The effect of adding the alkaline salts, covering a pH range of 8 to 12, is a progressive increase in the mobility (and consequently in stability of dispersion) with increasing pH value, at least to a maximum. The decrease following the maximum in those cases where maxima were observed is attributed by them to the same factor as obtains when

[23] Powney, J., and Wood, L. J., *Trans. Faraday Soc.*, **36**, 57–63 (1940).

neutral sodium chloride is the added salt, namely, the influence of the sodium ion.

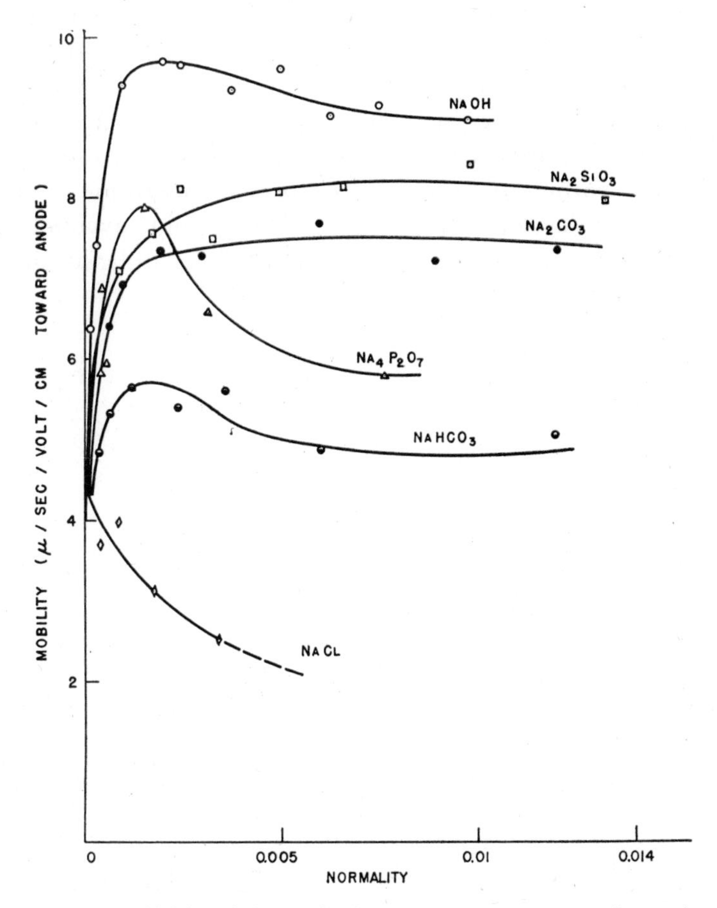

Figure 6–3. Mobility of dispersed oil droplets vs. concentration of several builders in the continuous phase.[23]

Additional Properties of Builder Solutions

Aqueous builder solutions have several important properties in addition to those discussed above, some of which may even dictate the final choice of one builder in preference to another. The relative extents of attack of different builders on fabrics, as measured by loss in fabric strength, and the bleaching action of some builders must be considered. Probably the most important of these additional properties from the standpoint of aiding the detergent process is the ability of several of the builders to "soften" the wash water, usually by sequestering the calcium and magnesium ions present in hard water. Such action is related to detergency in the sense that it prevents the deposition of insoluble soaps on the fabric and it disposes of polyvalent cations which might otherwise adversely affect the stability of dispersions. However, the general processes of water softening or sequestering of "hardness" are only indirectly related to detergency and are not within the scope of the present discussion. Certainly, with the highly developed methods of water softening available today, the use of hard water for commercial laundering can be justified only under unusual circumstances.

7

Electrical Phenomena in Detergency

General

The micelles present in aqueous colloidal systems are either positively or negatively charged, the charge ordinarily being considered to arise from one or more of the following conditions:

(a) Ionization of the material comprising the micelle;
(b) Capture of an ion by the micelle from the surrounding medium;
(c) Electrification by contact.

An example of case (a) would be the charged soap micelles, as discussed in Chapter 3, whether the ionization takes place before or after aggregation into the micelle. It is true that, in the water surrounding a charged micelle, there is an abundance of oppositely charged ions (gegenions), but none of these gegenions is sufficiently associated with the micelle to fully satisfy its charge. The result is a net charge on the micelle.

An example of case (b), above, would be the capture of a hydroxyl ion by, say, clay hydrosol, as follows:

$$\underset{\text{(neutral clay micelle)}}{(Al_2O_3\cdot 2SiO_2\cdot 2H_2O)_x} + (OH)^- \rightarrow \underset{\text{(negatively charged micelle)}}{[(Al_2O_3\cdot 2SiO_2\cdot 2H_2O)_x\cdot (OH)]^-} \quad (1)$$

The manner of capture of the hydroxyl ion by the clay micelle in this case is a function of adsorption.

The phenomenon of electrification by contact, as in case (c), above, involves the stripping of electrons from the atoms of one body when "rubbed" by a second body and the taking up of these electrons by the atoms of the second body. The production of electrostatic charge in such operations as the rubbing of amber with cloth and the combing of hair with a hard-rubber comb are well-known demonstrations of this phenomenon. The body which takes up the electrons becomes negatively charged and the body which loses them becomes positively charged. One requisite for producing such charges is the presence of a surrounding medium which has low conductivity. Fabrics such as cotton, wool, and silk can become highly charged in air when rubbed with a suitable body because air is a poor conductor. Rubbing of these fabrics in a nonconducting liquid medium may also produce electrostatic charge, as is evidenced by the dangers of electrostatic sparking in dry-cleaning operations. On the other hand, in laundry washing operations where the detergent solution is relatively highly conductive, contact electrification of either the fabrics or of soil particles should not be significant. Under such conditions the detergent solution serves as a conductor to effect the equal distribution of electrons throughout the system and thus prevent the accumulation of localized positive and negative static charges.

Michaelis[1] has advanced a fourth possible mechanism for the attainment of a charge in water by colloidal material, particularly cellulose, hydrocarbons, and other materials which have no great tendency to dissociate and a relatively slight tendency to oriented adsorption. Michaelis assumes a selective adsorption of hydroxyl ions (OH^-) from the water phase because of the capillary (surface) activity of this ion, so that there will exist near the surface of the nonaqueous phase a preponderance of negatively charged ions. It is questionable whether this proposed mechanism is truly a separate case or is a special application of the previously discussed mechanism involving the capture of an ion from the surrounding medium; however, it does afford a rather simple explanation of the attainment of a charge by presumably inert material.

As pointed out by Gortner,[2] most hydrosols contain negatively charged micelles. Micelles are usually negative when in contact with a liquid having a high dielectric constant, and positive when in contact with a liquid of low dielectric constant. The dielectric constant of water is 81, which is very high in comparison to that of most liquids.

---

[1] Michaelis, L., "The Effects of Ions in Colloidal Systems," Baltimore, Williams and Wilkins Company, 1925.

[2] Gortner, R. A., "Outlines of Biochemistry," New York, John Wiley & Sons, Inc., 1929.

As a general rule only the oxides and hydroxides of metals and certain basic organic compounds are positively charged in water.

The foregoing discussion of the source of charge on micelles is equally applicable to the charge on fiber surfaces. The manner in which fibers may become electrified by contact has been previously mentioned. Also, as chemical substances they may lose ions from or take up ions on their surfaces in such a way that their surfaces become charged. In this connection, it is surface charge in which we are interested. If we consider the interior of the fiber to be fundamentally a continuous homogeneous phase, electrical charges can exist only on the surface of the fiber and not on the interior. Electrical charge, like so many other phenomena, is a function of heterogeneity. In the present instance the heterogeneity is the localized difference in composition at the foci on the surface where ions or electrons are lost or adsorbed.

As is to be expected, the charged surface of a fiber or a colloidal micelle in water has ions of the opposite sign attracted to it, to form two layers of oppositely charged electricity as in a condenser. Helmholtz[3] originally advanced this concept of an electrical double layer

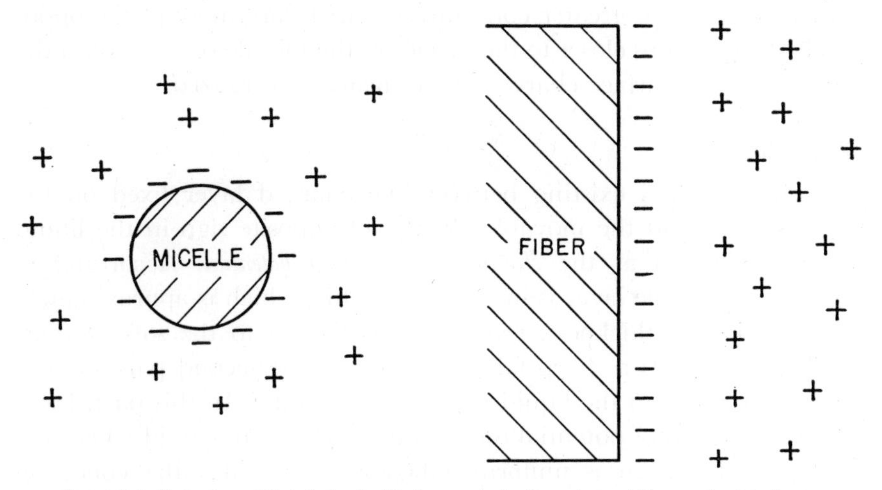

Figure 7–1. Illustration of Helmholtz double layers at micelle-water and fiber-water interfaces.

at surfaces and it is to this double layer that many of the properties of colloid systems can be attributed. Formation of the Helmholtz double layer can be pictured as illustrated in Figure 7–1. In this

[3] Helmholtz, H., *Wied. Ann.*, **7**, 337–81 (1879).

figure, the micelle or the fiber surface, as the case might be, negatively charged possibly by adsorption of negative ions from the surrounding medium, holds under its influence a layer of positively charged gegenions in the surrounding medium. While these gegenions cannot be considered chemically combined with the surface, their freedom of movement through the surrounding medium is definitely restrained by the negative surface charges. These gegenions may be continually replaced by like ions from farther out in the medium; however, the net result is a more or less average constant layer as pictured.

If the electrical double layer is to be considered as a condenser, then the distance between the two layers is of interest in determining the capacity of the condenser. Helmholtz has been interpreted erroneously as stating that this distance is the diameter of one molecule. Gouy[4] has calculated the distance to be 0.96, 9.6, and 1010 millimicrons in 0.10N NaCl, 0.001N NaCl, and pure water, respectively. Thus, the concentration of the ion of charge opposite in sign to the surface charge determines the distance between the two oppositely charged layers. The greater the concentration the less is the distance between layers and the lower is the capacity of the condenser. In other words, with greater concentration there are more of the oppositely charged ions closer to the surface of the micelle or fiber, with the result that the surface charge is more nearly neutralized.

Zeta Potential

The potential existing between the charged layer fixed on the solid surface and the movable charges of opposite sign in the liquid phase is known as the *electrokinetic* or *zeta potential*. According to Stern,[5] the double layer is in two parts. One, which is approximately a single ion in thickness, remains almost fixed to the solid surface and in it there is a sharp fall of potential. The second part extends some distance into the liquid phase and is diffuse. In this part, there is a gradual fall of potential out into the body of the liquid where the charge distribution is uniform. Glasstone[6] illustrates the condition at the solid-liquid boundary as shown on Figure 7–2, where the shaded portion represents the solid and the vertical broken line the

[4] Gouy, M., *J. phys.*, (4) **9**, 457–67 (1910).

[5] Stern, O., *Z. Elektrochem.*, **30**, 508–16 (1924).

[6] Glasstone, S., "Textbook of Physical Chemistry," 2nd ed., New York, D. Van Nostrand Company, Inc., 1946.

extent of the fixed part of the double layer. "If the potential of the solid is indicated by A and that of the bulk of the liquid by B, the fall of potential in between may occur in two ways (Figure 7–2, 1 and 2), depending on the characteristics of the ions or molecules present in solution which make up the outer portion of the fixed layer. In each case AC is the sharp fall of potential in the fixed part and CB the gradual change in the diffuse part of the double layer." The potential marked ζ in each diagram is the zeta potential and is the potential between the fixed and the movable layers. Glasstone further points out "the distinction between the electrokinetic potential and the

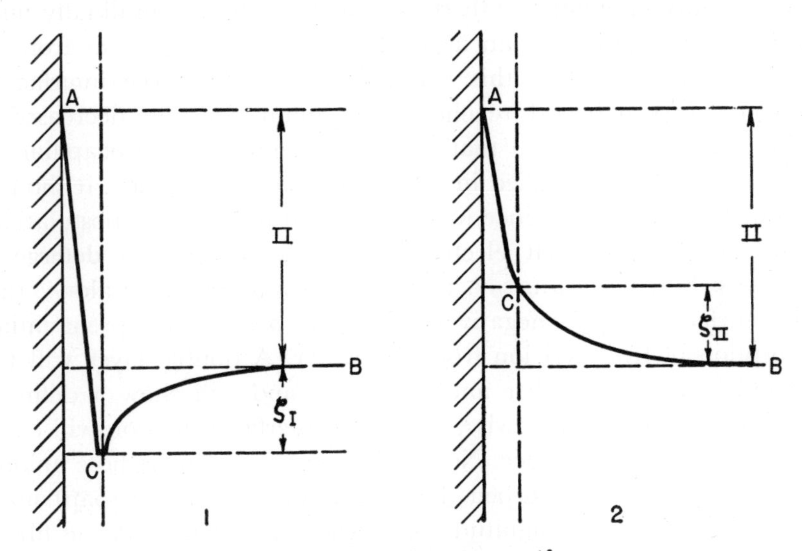

Figure 7–2. Electrokinetic potential.[6]

thermodynamic, or reversible, potential existing between the solid and the bulk of the solution; the latter is represented by (II in Figure 7–2, 1 and 2). The two quantities are quite independent and must not be confused." Freundlich[7] shows that substances such as electrolytes may have a marked effect on the zeta potential but that the thermodynamic potential is hardly affected by them.

Several properties of colloids and of solid-liquid interfaces in general may be attributed to zeta potential; namely, electrophoresis, electroendosmosis, streaming potential, and sedimentation potential.

---

[7] Freundlich, H., "Colloid and Capillary Chemistry," translated by H. S. Hatfield, New York, E. P. Dutton and Company, 1926.

These properties are simply manifestations of the existence of zeta potential rather than measures or indices of detergent action that might take place in the systems. It is true, however, that they serve as means of measuring the zeta potential. The significance of zeta potential in detergency is discussed elsewhere (Chapters 5, 6, and 9).

Electrocapillarity

The effect of an electric charge on interfacial tension was first studied extensively in 1873 by Lippmann,[8] using mercury-aqueous solution interfaces. He found that the mercury surface, which initially had a positive potential with respect to the solution, gradually was made more negative by an applied electromotive force (as from a battery) and that with this transition in potential the interfacial tension at first increased, reached a maximum, and then decreased.

While it is not known that the phenomenon of electrocapillarity has been previously studied in connection with detergency, it should be worth while to introduce into the present discussion a postulation of the significance that electrocapillarity may have in detergent processes. Starting with, say, a fiber immersed in water alone, the fiber surface will be negatively charged, perhaps by preferential adsorption of hydroxyl ion from the water. A double layer will be created between the fiber and the water and will show a definite potential (negative). Likewise, a definite interfacial tension will exist at the fiber-water interface. Now, an electrolytic solute whose anions are more positively adsorbed than its cations (such as a soap) may tend to increase the magnitude of the negative charge on the fiber, with consequent increase in the negative double-layer potential. At least part of the lowering of the interfacial tension attending positive adsorption of the anions may be attributable to the increased potential and the corresponding decreased work required to increase the area of the interface.

Based on studies by Kopaczewski and Rosnowski[9,10] of the penetration of colloidal material into capillary interstices, other functions of electrocapillarity in detergency are indicated as follows:

(a) Penetration of negatively charged particles into interstices of cellulose immersed in water (wherein the cellulose surfaces are

[8] Lippmann, G., *Ann. Phys.*, **149**, 546 (1873).
[9] Kopaczewski, W., and Rosnowski, M., *Compt. rend.*, **185**, 450–3 (1927).
[10] Kopaczewski, W., and Rosnowski, M., *Protoplasma*, **5**, 14–34 (1928).

negatively charged) may be reduced or even prevented by reduction or reversal of the charge on the cellulose.

(b) Electrolyte anions often have a pronounced effect in aiding the penetration of negatively charged particles into negatively charged interstices, whereas cations may effectively block such penetration.

Such electrocapillary effects as these may have a bearing both on the way in which cloth is soiled and on the way in which the soil is removed during washing.

8

Influence of Builders on the Nature and Surface Activity of Detergent Solutions

Most of the effectiveness of builders in detergency is due to their influence on the properties and actions of soap and synthetic detergent solutions. For some time builders were considered to be merely soap extenders or "diluents," relatively inexpensive materials whose presence did not noticeably reduce the efficacy of the soap but permitted a greater bulk of product at lesser cost. However, it is now known that properly chosen and properly used builders in conjunction with soaps or other detergents produce combinations whose detergent efficiency is definitely superior to that of solutions of detergent alone.

Builders and Dissociation of Detergents

Again, as in the chapter on the properties of unbuilt detergent solutions, the first consideration is the dissociation and hydrolysis of soaps and synthetic detergents, and particularly the influence thereon of builders.

Being quite strong electrolytes, all ionic detergents should be completely dissociated when in solutions of sufficient dilution. Departure of dilute ionic detergent solutions from the behavior of ideal solutions should be attributable to interionic attraction, in accordance with the Debye-Hückel theory. In other words, in the space immediately surrounding any ion in the solution there will be a greater number of ions of opposite charge than of like charge, with the result that the particular ion is no longer free to act as an independent entity. The higher the concentration the less will be the freedom of any one ion in the solution. At concentrations sufficiently high (perhaps in the range of significant formation of micelles), at least part of the detergent should be present in solution in a form which, for all practical purposes, may be considered as molecular.

In the light of the foregoing discussion, an equation such as equation (2) of Chapter 3,

$$\mathrm{RCOONa} \rightleftharpoons \mathrm{Na^+} + \mathrm{RCOO^-} \tag{1}$$

must carry with it certain reservations. In a strict sense, the expression RCOONa holds only at concentrations where the molecular species may be considered to exist. Below this concentration, only the ions Na^+ and $RCOO^-$ will exist and their ability to act as independent entities is a function of such factors as concentration, temperature, and composition of the R of the $RCOO^-$.

The extent to which the properties of any species of ions in solution are a function of the concentration of the particular species may be expressed as the activity coefficient of the species. In highly dilute solutions, where attraction between oppositely charged ions exerts only a slight influence, the activity of any ionic species may be considered equal to its concentration. However, this relationship no longer holds when interionic attraction plays a significant role. The activity coefficient for any species of ions of strong electrolytes in dilute solution, as a function of concentration and temperature, has been developed quantitatively by Debye and Hückel,[1] who derive the expression

$$-\ln \gamma_i = \frac{e^3 z_i^2}{(DkT)^{3/2}} \sqrt{\frac{2\pi N\mu}{1000}} \tag{2}$$

In this equation[2]

γ_i = activity coefficient of ion species i,
z_i = valence of ion species i,
e = charge of an electron = 4.803×10^{-10} electrostatic units,

---

[1] Debye, P., and Hückel, E., *Physik. Z.*, **24**, 185–206, 305–25 (1923).

[2] An excellent derivation of the Debye-Hückel equation is given by Getman, F. H., and Daniels, F., "Outlines of Physical Chemistry," 7th ed., pp. 667–70, New York, John Wiley and Sons, Inc., 1943.

D = dielectric constant of the solution = 78.56 for water at 298°K,
N = Avogadro's number = 6.023×10^{23},
k = gas constant per molecule = $1.3805 \times {}^{-16}$ erg per degree,
μ = ionic strength = one-half the sum of the products of the concentration of each ion species in the solution times the square of its valence = $\frac{1}{2}(c_1 z_1^2 + c_2 z_2^2 + c_3 z_3^2 \ldots)$, and
c_i = concentration of ion species i in moles per liter.

It may be seen from the right-hand side of equation (2) that, for any given ion species in solution, its activity coefficient is dependent on the variables, temperature and ionic strength. In turn, the ionic strength of the solution is dependent on the respective concentrations and valences of all the ion species in the solution.

If we consider equation (1) as representative of dissociation conditions obtaining at equilibrium in a soap solution of such concentration that soap molecules may be considered to exist, then the effect of a builder when added to the solution is a "mass" action driving the equilibrium to the left. However, this is simply a specific case which may be fitted to equation (2) as readily as more dilute solutions. In words, equation (2) tells us that the presence of additional ions from a builder increases the ionic strength of the solution and correspondingly reduces the activity coefficient of, say, the fatty anions already in solution.

The quantitative application of equation (2) can be readily demonstrated using, say, sodium lauryl sulfate and sodium sulfate builder for an illustration:

Assume an unbuilt sodium lauryl sulfate solution of $0.1M$ concentration at 27°C. On the basis of complete dissociation, the concentration of univalent Na^+ ions will be $0.1M$ as will be the concentration of the univalent lauryl sulfate anions. The ionic strength of the solution is then

$$\mu = \frac{1}{2}(c_{Na^+}(1)^2 + c_{R^-}(1)^2) = \frac{1}{2}(0.1 + 0.1) = 0.1$$

where subscript R^- denotes lauryl sulfate anion. Substituting into equation (2) and solving for γ_{R^-} then gives

$$\gamma_{R^-} = 0.693$$

where γ_{R^-} is the activity coefficient of the lauryl sulfate anions in unbuilt $0.1M$ solution.

If we now make the sodium lauryl sulfate solution $0.05M$ with respect to sodium sulfate and assume complete ionization of the sodium sulfate, the ionic concentrations in the solution are now

$$\begin{aligned} c_{Na^+} &= 0.1M \text{ (from the detergent)} + 0.1M \text{ (from the builder)} \\ &= 0.2M \\ c_{R^-} &= 0.1M \\ c_{SO_4^=} &= 0.05M \end{aligned}$$

The ionic strength of the solution is then

$$\begin{aligned} \mu &= \tfrac{1}{2}(c_{Na^+}(1)^2 + c_{R^-}(1)^2 + c_{SO_4^=}(2)^2) \\ &= 0.25 \end{aligned}$$

Substitution into equation (2) as before now gives us a value

$$\gamma_{R^-} = 0.556$$

The value for γ_{R^-} in unbuilt $0.1M$ solution indicates that the activity of the lauryl sulfate anions (that is, their ability to act as independent entities) is only 0.693 of what it would be if the anions were entirely away from the influence of the Na^+ cations. Upon addition of $0.05M$ sodium sulfate the activity of the lauryl sulfate anions is further reduced so that it is then only 0.556 of what it would be in the absence of interionic attraction. In view of the fact that the activity coefficient of, say, fatty anions is a measure of the ability of those anions to enter into any thermodynamic process, the above calculations illustrate in a general way one of the effects which may be expected from the addition of electrolyte builder to an ionic detergent solution.

Builders and Hydrolysis of Soaps

Reference is made to equation (3) of Chapter 3 for hydrolysis in a very dilute soap solution,

$$RCOO^- + H_2O \rightleftharpoons RCOOH + OH^- \qquad (3)$$

Further, Ekwall's equations for the formation of acid soap in somewhat more concentrated solution (equations (5) and (6) of Chapter 3) may be expressed by the general equation

$$RCOOH + RCOO^- \rightleftharpoons (H(RCOO)_2)^- \overset{Na^+}{\rightleftharpoons} RCOOH \cdot RCOONa \qquad (4)$$

Examination of these equations indicates that the mass action of any added electrolyte builder not only decreases the activity of $RCOO^-$

anions as discussed above in connection with dissociation but also concomitantly inhibits the formation of fatty acid or of acid soap. In the case of alkaline builders, the hydroxyl ions resulting from hydrolysis of the builder may be considered to exert more than a mass-action effect on equations (3) and (4): they chemically suppress the formation of fatty acid and of acid soap.

It is well, for the case of soap solutions, to consider dissociation and hydrolysis together, rather than as separate reactions. Based on

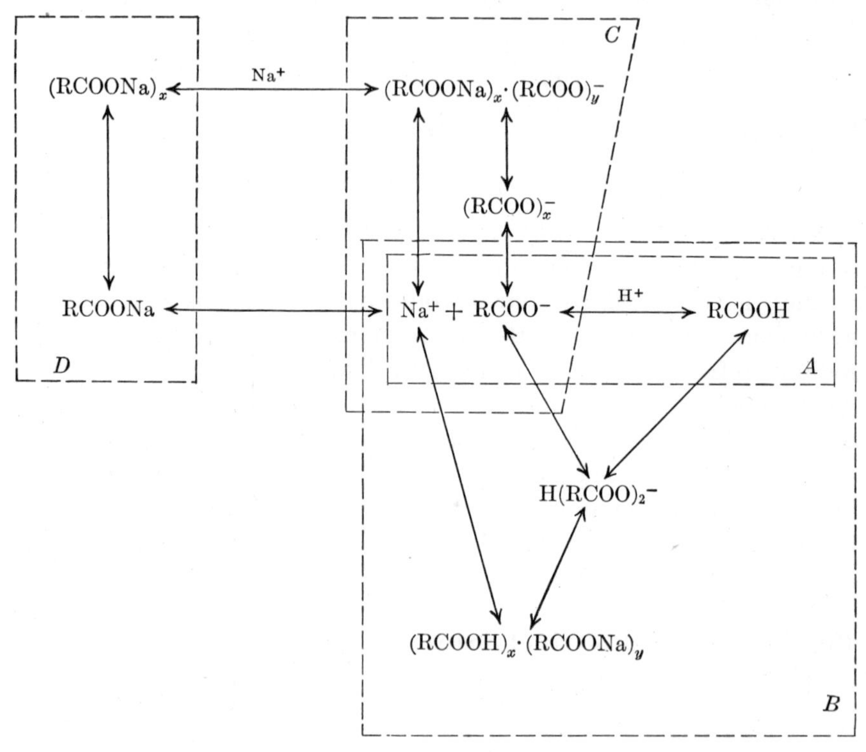

Figure 8–1. Schematic illustration of dissociation and hydrolysis in soap solutions.

the summary in Chapter 3 (page 38) relative to the composition of soap solutions at different concentrations, we may write as a "schematic" representation of the combined reaction at any concentration the equation shown in Figure 8–1. The reactions within the areas designated *D* and *A* represent those which are predominant at high and very low concentrations, respectively. Thus, at concentrations in the range of 5 per cent the predominant reaction very well may be the formation of molecular soap micelles, with little dissociation or hydrolysis, whereas at concentrations below the limiting solubility of the fatty acid formed from hydrolysis the predominant reactions will be complete dissociation and quite appreciable hydrolysis. Starting with a very dilute solution, by progressively increasing the soap concentration the predominant reactions shift as in going from A to B to C to D.

A mathematical treatment of the various equilibria shown in Figure 8–1, by the use of dissociation and equilibrium constants, is difficult, as can be illustrated by the following:

If we consider the soap molecule as the starting material the first constant with which we are concerned is that for dissociation. This constant may be designated K_1 and may be expressed as

$$K_1 = \frac{c_{Na^+} \times c_{RCOO^-}}{c_{RCOONa}} \tag{5}$$

where c designates the equilibrium concentration of the particular component shown as a subscript. However, in the entire range of lower concentrations where dissociation of the molecules is complete, K_1 has no finite value.

For the reaction in area A, Figure 8–1, the hydrolysis constant at equilibrium, K_2, might be expressed as

$$K_2 = \frac{c_{RCOOH}}{c_{H^+} \times c_{RCOO^-}} \tag{6}$$

However, the influence of the Na^+ ion on the activity of the $RCOO^-$ ion and, thus, on the ability of the latter ion to enter into the reaction cannot be overlooked.

Going next to the case of the reactions in area B, our mathematical treatment becomes more difficult. The equilibrium constant in this case may be expressed as

$$K_3 = \frac{c_{RCOOH} \times c_{(H(RCOO)_2)^-} \times c_{(RCOOH)_x \cdot (RCOONa)_y}}{c_{Na^+} \times c_{RCOO^-} \times c_{H^+}} \tag{7}$$

wherein the exponents to which the values in the denominator should be raised have been omitted for simplicity. However, such a constant is worthless, if for no other reason than the impossibility of its measurement. In the first place, such a formula as $(RCOOH)_x \cdot (RCOONa)_y$ is only a general expression wherein x and y are functions of concentration. The same probably holds also for the $H(RCOO)_2^-$. However, based on relative *activities*, we may arrive at a statement which *qualitatively* describes the conditions in this area: So long as the activity of each of the $RCOO^-$ ions is high (as in the range of very low concentrations) the value of K_3 will be low; as the activity of each of the $RCOO^-$ ions decreases (with increased over-all concentration) there will be a greater tendency, first, to the formation of fatty acid and, then, to the formation of acid soap, with corresponding increase in K_3.

Mathematical treatment of the equilibria in areas *C* and *D* of Figure 8–1 involves the same difficulties as those just discussed. In each case, the extent of micellar aggregation, whether ionic, molecular, or a combination of the two, is a function of concentration.

The mathematical treatment attempted above, while leaving much to be desired in the way of quantitative information, nevertheless illustrates qualitatively the effects of over-all concentration on the various equilibria. It follows that the effects thus produced by changes in over-all concentration are quite generally similar to the effects produced by added electrolyte builder. The addition of builder to a soap solution of any given concentration has an effect on the particular *K* value comparable to that produced simply by an increase in soap concentration.

One further point to be borne in mind in connection with the suppression of hydrolysis of soaps by means of builders is that, to attain such suppression, the pH of the builder at the concentration used must be higher than that of the unbuilt soap solution. Thus, in the case of a soap solution having a pH of, say, 10.4, use of sodium bicarbonate or borax as a builder to suppress hydrolysis would be ineffective because of their tendency to lower the pH. In fact, such use would be comparable to the use of an acid and would favor more extensive hydrolysis. A common contention that modified soda acts also to lower the pH of soap solutions has been at least partially refuted by Bacon, Hensley, and Vaughn,[3] who have shown that, at a temperature of 60°C, the pH of soap solutions is practically unaffected and at 80°C is raised by modified soda.

Builders and Surface Tension

The general effect of electrolyte builders on the surface tension of detergent solutions is, with some qualification, an increase in ten-

[3] Bacon, L. R., Hensley, J. W., and Vaughn, T. H., *Ind. Eng. Chem.*, 33, 723–30 (1941).

sion for soap solutions and, in contrast, a decrease for solutions of unhydrolyzed anionic detergents.

Starting with an unbuilt *soap* solution, as illustrated in Figure 8–2,[4] the addition of alkaline builder in progressively greater amounts has the effect of first raising the surface tension of the solution to a maximum and then reducing the tension, but not to the original lower value of the unbuilt solution. Soap solutions buffered to increasing pH values by alkaline salts so chosen that the salt concentration is constant show surface-tension values which increase to a maximum and then remain quite constant. This effect is illustrated by

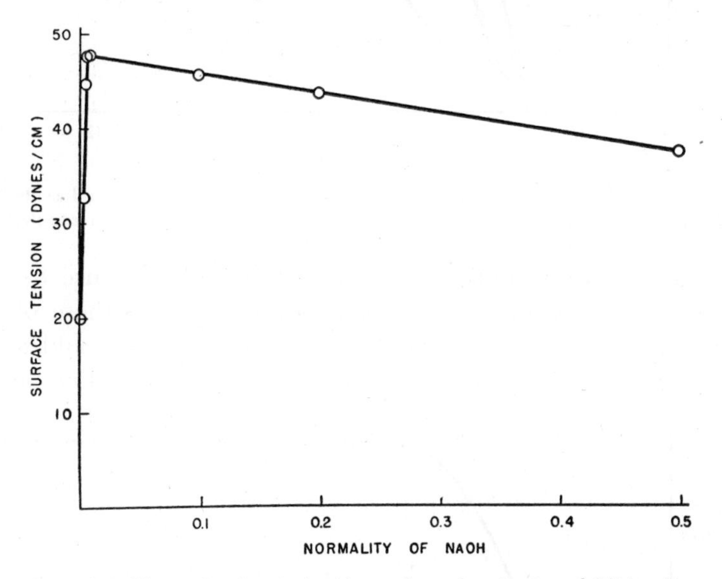

Figure 8–2. Effects of sodium hydroxide on the surface tension of 0.1*M* sodium nonylate solution.[4]

Figures 8–3 and 8–4, from the work of Long, Nutting, and Harkins.[5] Long points out that the lowering of the surface tension as the buffer concentration increases (as in going from curve 3 to curve 1 in Figure 8–3) is an effect due to the concentration of added salt rather than an effect due to pH. Confirmation of this was obtained by adding to a solution of a given pH, and 0.001*N* or 0.01*N* with respect to buffer, sufficient neutral sodium chloride to make the total concentration of

[4] Harkins, W. D., and Clark, G. L., *J. Am. Chem. Soc.*, **47**, 1854–6 (1925).

[5] Long, F. A., Nutting, G. C., and Harkins, W. D., *J. Am. Chem. Soc.*, **59**, 2197–2203 (1937).

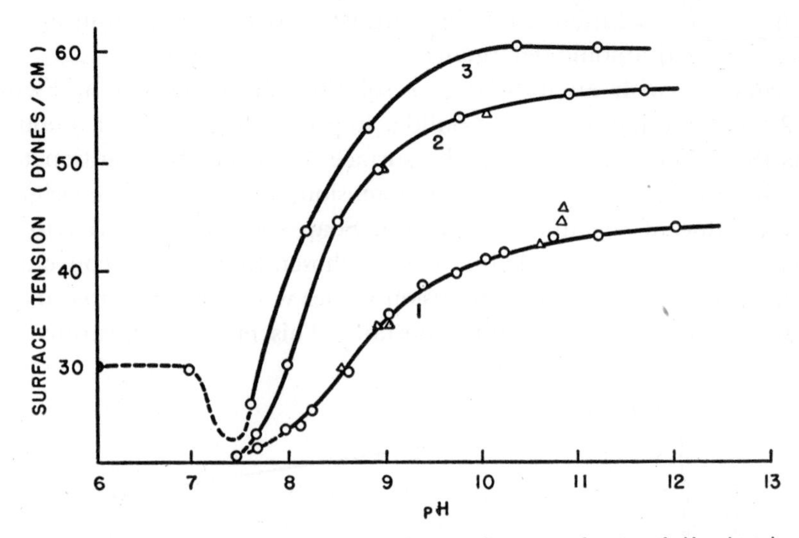

Figure 8–3. Surface tension of 0.005M sodium laurate solutions as a function of pH and total salt concentration.[5] Curves 1, 2 and 3 are for 0.1N, 0.01N and 0.001N buffers, respectively.

added salt 0.1N. As indicated by the triangles in Figure 8–3, the surface-tension values thus obtained corresponded to those of solutions 0.1N with respect to buffer alone. From this it would appear that increase in surface tension of soap solutions by alkaline builders

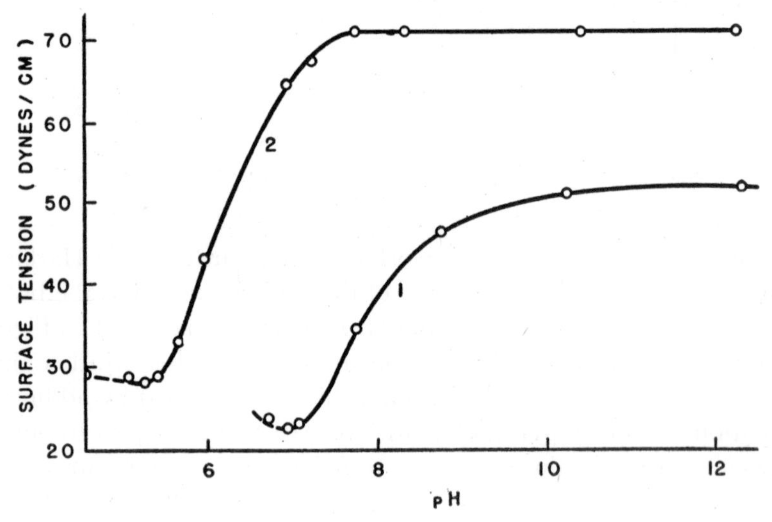

Figure 8–4. Surface tension of sodium nonylate solutions as a function of pH.[5] Curves 1 and 2 are for 0.1M and 0.005M solutions of sodium nonylate, respectively. Both series of solutions 0.01N with respect to buffer.

is a function primarily of hydroxyl-ion concentration. Opposing this increase in the cases just cited is the "mass" effect of the total concentration of added salt at a given pH; the greater this concentration the less is the net increase in surface tension effected by the hydroxyl ions.

If, as suggested in Chapter 4 (page 72), fatty acid from hydrolysis is a particularly significant contributor to the lowering of the surface tension of dilute soap solutions, with fatty ion next in order, we may reason that the *primary* effect of increasing the hydroxyl-ion concentration is a suppression of formation of fatty acid (to account for increased surface tension). On the other hand, we also may reason that an increase in total concentration of added salt likewise should

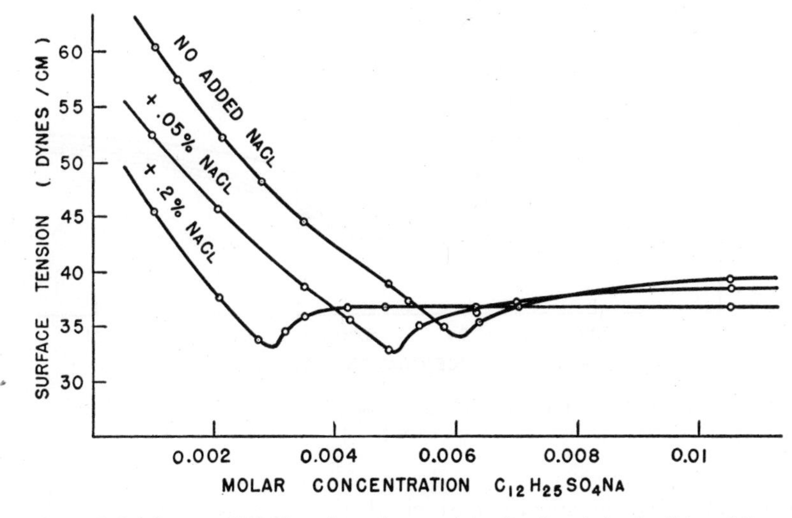

Figure 8–5. Influence of NaCl on the surface tension of sodium dodecyl sulfate solutions at 20°C.[6]

suppress formation of fatty acid or decrease the activity of the fatty ions. It would appear, then, that suppression of activity and hydrolysis cannot alone account for the net combined effect of hydroxyl-ion concentration and total salt concentration on the surface tension of soap solutions. The possibility that the net effect is a manifestation of another action by the builder will be discussed later in connection with the effect of builders on micelle formation.

The effect of electrolytes on the surface tension of anionic synthetic detergents is illustrated by the data in Figures 8–5 and 8–6, from the work of Powney and Addison[6] with sodium dodecyl sulfate solutions. These data show that, when the dodecyl sulfate concentrations are below the critical value for minimum surface tension, the addition of electrolyte may markedly reduce the surface tension of the solutions. Further, by the addition of electrolyte the concentration of dodecyl sulfate necessary to attain minimum surface tension may be reduced considerably. Finally, this effect of the electrolyte on anionic detergents is a definite function of the valence of the added

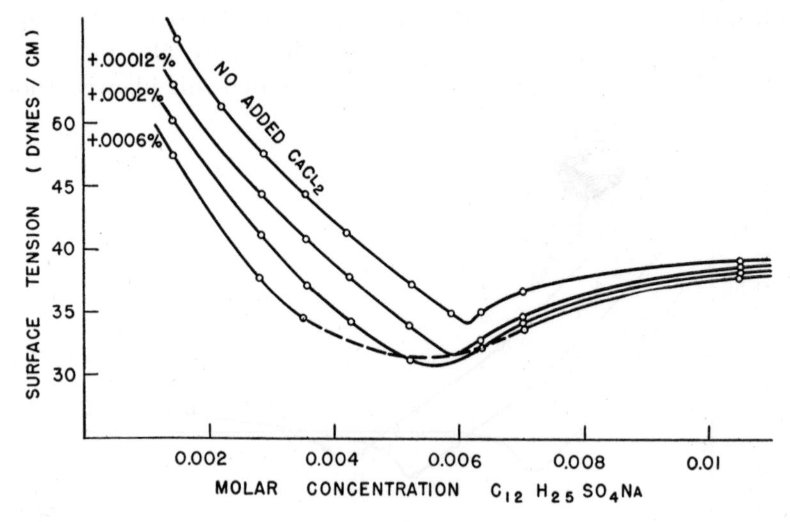

Figure 8–6. Influence of $CaCl_2$ on the surface tension of sodium dodecyl sulfate solutions at 20°C.[6]

cation. Thus, in the specific case cited, the ratio between the amounts of sodium chloride and of calcium chloride to bring about a given lowering of surface tension (when added to sodium dodecyl sulfate solutions below the critical concentration) is of the order of 200:1. A discussion of the probable way in which this valence effect comes about will be given later (page 156).

Harris,[7] in a review of electrolyte builders for surface-active agents, has illustrated well the extent of the above-discussed influence of builders on the surface activity of solutions of anionic synthetic

---

[6] Powney, J., and Addison, C. C., *Trans. Faraday Soc.*, **33**, 1253–60 (1937).
[7] Harris, J. C., *Am. Dyestuff Reptr.*, **37**, 266–70 (1948).

detergents. As shown in Table 8–1 (from various sources) the amount of surface-active agent can be reduced by as much as $\frac{1}{10}$ to $\frac{1}{2}$ by the use of electrolyte to accomplish the same surface- or interfacial-

TABLE 8–1. EFFECTS OF ELECTROLYTES ON SURFACE AND INTERFACIAL TENSIONS OF SOLUTIONS OF SURFACE-ACTIVE AGENTS [7]

Agent and Electrolyte	Concentration	Effect of Added Electrolyte	Author Reference
Sodium cetyl sulfate	0.0001*M*	Surface tension reduced by	5
NaCl	.02*M*	24.6 dynes/cm.	
HCl	.02*M*	25.2 dynes/cm.	
NaOH	.02*M*	23.4 dynes/cm.	
Sodium dodecyl sulfate	.0008*M*		
NaCl	1%	1/9 active to produce same interfacial tension	6
Sodium dodecyl sulfate	0.0025*M*		
NaCl	.2%	1/2 active to produce same surface tension	6
Sodium dodecyl sulfate	0.004*M*		
$CaCl_2$	00045%	1/2 active to produce same interfacial tension	6
Sodium dodecyl sulfate	ʋ.ᴊᴊ37*M*		
$CaCl_2$	.0006%	1/2 active to produce same surface tension	6
Igepon-T	0.005%		
($C_{17}H_{33}CON(CH_3)C_2H_4SO_3Na$)		Interfacial tension reduced	
NaCl	.32*N*	from 6–25 to 1 dyne/cm.	8
$CaCl_2$	.003*N*		
$LaCl_3$	.00005*N*		
Sodium tridecane-7-sulfonate	0.00227*M*		
	(.0687%)		
NaCl	.13*N*	1/6 active to produce same	
KCl	.1*N*	surface tension	9
$MgCl_2$	.01*N*		
$CaCl_2$	.01*N*		

tension lowering as that of the unbuilt solutions. Harris[10] has also studied the effects of several electrolytes on the surface and interfacial tensions of solutions of dodecyl benzene sodium sulfonate, with

---

[8] Robinson, C., *Nature,* **139,** 626 (1937).

[9] Dreger, E. E., Keim, G. I., Miles, G. D., Shedlovsky, L., and Ross, J., *Ind. Eng. Chem.*, **36,** 610–7 (1944).

[10] Harris, J. C., *Oil and Soap,* **23,** 101–10 (1946).

the results shown in Table 8–2. In each concentration, except for the unbuilt solutions shown on the first line, 60 per cent of the total represents the indicated builder and only 40 per cent is the active agent. In other words, by the use of builders, it is possible to use only

TABLE 8–2. EFFECT OF CATION UPON SURFACE AND INTERFACIAL TENSION OF DODECYL BENZENE SODIUM SULFONATE [10]

Builder	Surface Tension			Interfacial Tension (Nujol)		
	1.0% soln.	0.25% soln.	0.0625% soln.	1.0% soln.	0.25% soln.	0.0625% soln.
None	32.0	31.2	30.3	5.2	3.5	3.1
NaCl	28.4	29.7	31.6	2.5	2.5	3.5
Na_2SO_4	29.2	30.2	30.4	2.9	2.8	3.3
Na_2CO_3	28.5	30.3	30.9	2.7	3.6	5.1
$Na_4P_2O_7$	29.4	30.6	34.8	2.7	4.3	8.4

Note: The total solute for all built solutions consisted of 60 per cent builder and 40 per cent active agent.

40 per cent as much active agent to attain surface and interfacial tensions which are, in most cases, lower than those of the unbuilt solutions.

An effect on surface tension opposite to that produced by alkaline builders has been noted by Lottermoser and Tesch[11] and by Cupples[12] (see Figure 9–1, Chapter 9) in solutions of soap to which excesses of fatty acid have been added. Such additions displace the minimum of surface tension toward lower concentrations and suppress the increase in surface tension that is usually expected with higher soap concentrations.

A final significant effect of added salts on the surface tension of detergent solutions is the effect on the time required to attain equilibrium at the air-solution interface. Whereas dilute unbuilt detergent solutions may require considerable time to attain their equilibrium surface-tension values, Adam and Shute[13] have found that the presence of salts tends to "force" the surface tension to its final value almost at once. Nutting, Long, and Harkins,[14] in their study of the time effect on the surface tension of solutions of sodium cetyl and sodium lauryl sulfates, also determined the influence of added electrolytes on the rate of change of surface tension. As shown

[11] Lottermoser, A., and Tesch, W., *Kolloid-Beihefte*, **34**, 339–72 (1931).
[12] Cupples, H. L., *Ind. Eng. Chem.*, **29**, 924–6 (1937).
[13] Adam, N. K., and Shute, H. L., *Trans. Faraday Soc.*, **34**, 758–65 (1938).
[14] Nutting, G. C., Long, F. A., and Harkins, W. D., *J. Am. Chem. Soc.*, **62**, 1496–1504 (1940).

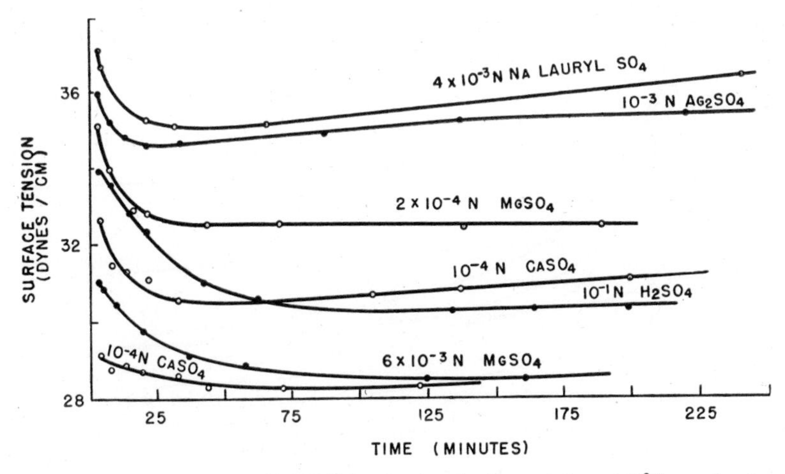

Figure 8–7. Surface tension of 4 × 10^{-3}N sodium lauryl sulfate solutions at 40°C, as a function of surface age, and the effect thereon of electrolytes.[14]

in Figures 8–7 to 8–10, taken from their data, a dual effect, previously noted by Adam and Shute, is produced by the electrolytes, namely, that of hastening the attainment of equilibrium and of producing surface tension *vs.* time curves similar to those of more concentrated solutions of the pure detergent.

The explanation by Doss[15] of the time effect on surface tension of unbuilt solutions, based on the existence of a potential barrier at

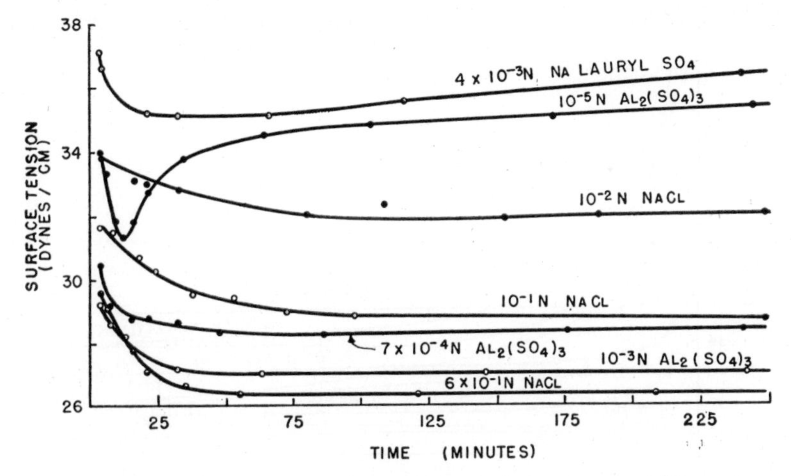

Figure 8–8. Surface tension of 4 × 10^{-3}N sodium lauryl sulfate solutions at 40°C, as a function of surface age, and the effect thereon of electrolytes.[14]

[15] Doss, K. S. G., *Kolloid-Z.*, **86**, 205–13 (1939).

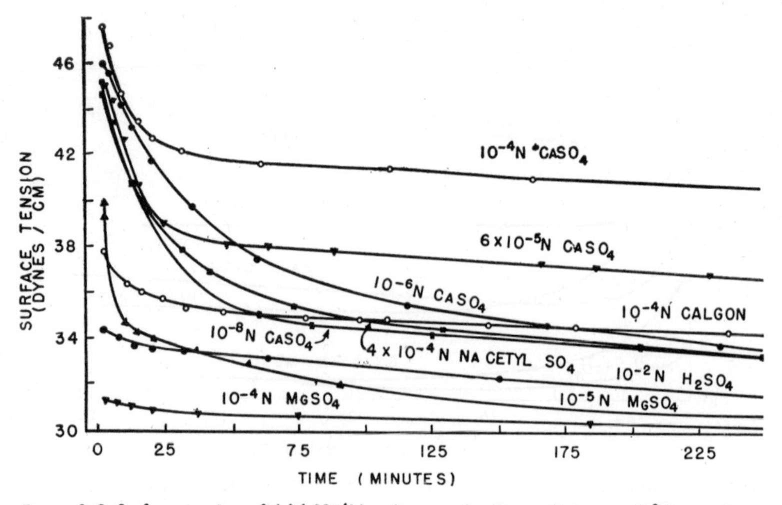

Figure 8–9. Surface tension of $4 \times 10^{-4}N$ sodium cetyl sulfate solutions at 40°C, as a function of surface age, and the effect thereon of electrolytes.[14]

the surface, has already been considered in Chapter 4 (page 57). The influence of added electrolytes on this time effect has been correlated by Nutting and coworkers with the appearance of micelles in the bulk of the solution. Adam and Shute consider that the mobility of an ionic micelle per ion is greater than that of simple ions. Nutting and coworkers consider that this greater mobility of micelles may play a significant part in the higher rate of attainment of surface-

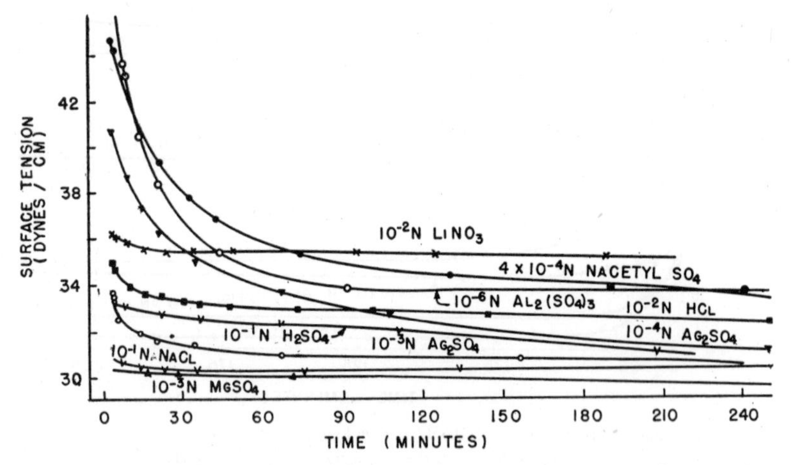

Figure 8–10. Surface tension of $4 \times 10^{-4}N$ sodium cetyl sulfate solutions at 40°C, as a function of surface age, and the effect thereon of electrolytes.[14]

tension equilibrium under conditions where micelles exist in the solution. Thus, by this postulation, the electrostatic barrier at the surface will still oppose entrance into the surface by micelles as it does by ions; however, the repulsive force *per ion* will be much less when the ions are associated in a micelle. Upon penetration into the surface layer, it must be assumed that the micelle "disintegrates" into the surface-active individual ions; this assumption is in line with the fact that the conditions causing aggregation within the bulk of the solution are no longer present once the aggregate reaches the surface.

Such a possible role for micelles as that postulated by Nutting in the *rate of attainment* of static surface tension must not be confused with the issue as to micelles being a significant surface-active species. In unbuilt detergent solutions where the detergent concentration is below the critical value for micelle formation, the addition of builder may so affect the solution that formation of micelles will occur. In effect, the builder may be considered to have decreased the "solubility" of the detergent or to have produced an effect similar to that existing in an unbuilt solution of a detergent of higher molecular weight. Unbuilt detergent solutions at the critical concentration show a very rapid rate of attainment of static surface tension and the addition of builder does not increase this rate. As the detergent concentration is decreased below the critical value, the rate in unbuilt solutions becomes slower and the concentration of a particular builder necessary to give rapid attainment of static surface tension becomes greater. At the same time, the observed surface tensions from the various detergent-builder combinations which give the critical concentration are not necessarily equal. If the micelle proper were the significant surface-active species, then we should expect quite equal surface tensions of solutions containing critical detergent-builder combinations. The conditions prompting micelle formation in relation to the conditions prompting maximum surface activity will be discussed later (page 160).

The above-discussed effects of builders on the surface tension of detergent solutions may be summarized as follows:

(a) Low concentrations of *alkaline* builders increase the surface tension of *soap* solutions of given concentration. The presence also of *neutral* builder decreases this effect of the alkaline builders.

(b) The addition of excess fatty acid to soap solutions produces the combined effect of displacing the minimum of surface tension in the direction of lower soap concentrations and of suppressing the increase in surface tension usually found when the soap concentration giving the minimum is exceeded.

(c) Neutral electrolyte builders show a pronounced effect in lowering the surface tension of solutions of unhydrolyzed anionic detergents. The extent of influence of these builders is a pronounced function of their cation valence, with the concentration of polyvalent cations necessary to attain a given lowering of surface tension being much lower than the corresponding concentration of univalent cations.

(d) It is both possible and practical (at least, with respect to surface-tension lowering) to build unhydrolyzed anionic detergents to the extent where the proportionate amount of builder may even predominate, and yet obtain surface tensions in many cases lower than those of unbuilt solutions of equal total solute concentration.

(e) The addition of electrolytes to solutions of unhydrolyzed anionic detergents hastens the attainment of equilibrium at the solution-air interface.

Builders and Interfacial Tension

The principal effect of electrolyte builders at the interface between detergent solutions and oils is quite generally a marked reduction in interfacial tension. In the absence of any appreciable surface activity of the builders themselves, this effect is evidence of a marked increase, by the presence of the builders, in the interfacial activity of the detergent.

The decrease in interfacial tension with increase in concentration of unbuffered alkalinity is illustrated by curves *A* and *B* of Figure 8–11, from the works of Shorter.[16] These curves may be contrasted with Harkins' and Clark's data for the *surface tensions* of 0.1*M* sodium nonylate solutions containing increasing amounts of sodium hydroxide (Figure 8–2). Curve *A* shows the results for varying amounts of alkali in 0.2 per cent soap solution and curve *B* for similar experiments with 0.5 per cent soap solution, all against pure benzene. In discussing his results, Shorter considered the most obvious explanation of the observed increase in drop number (decrease in interfacial tension), namely, the increase in amount of undecomposed soap by suppression

[16] Shorter, S. A., *J. Soc. Dyers Colourists*, **32**, 99–108 (1916).

of hydrolysis, to be quite inadequate. Thus, he states that "the only effect of the suppression of hydrolysis would be to make the concentration of the undecomposed soap in the solution equal to that in a somewhat stronger solution of soap containing no added alkali. It will be seen from Figure 8–11 that the drop number of the solution containing 2 grams per liter may be increased to more than 120 by the addition of a sufficient amount of alkali. Now experiment shows that a solution containing 20 grams of the soap per liter (and no added alkali) has a drop number of only 99. . . . It is evident, therefore, that some other effect, or effects, must operate in addition to the suppression of

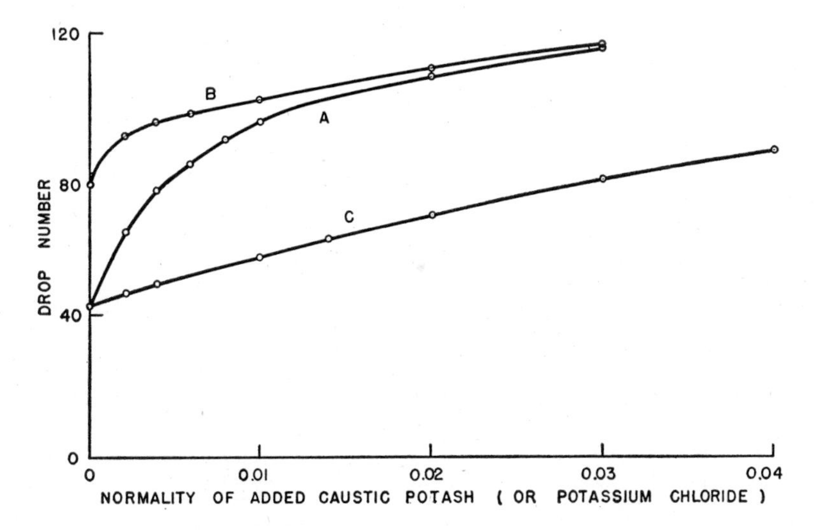

Figure 8–11. Effects of builder on the drop number of soap solutions against benzene (increase in drop number denotes decreased interfacial tension).[16]

hydrolysis." Shorter considered that this increase in surface activity of the soap might be due to the alkali increasing the colloidal nature of the soap, in a manner somewhat similar to the action of "salting out" of soap by electrolytes, and showed in support of this possibility the results of curve *C*, Figure 8–11, for 0.2 per cent soap solution and variable amounts of potassium chloride. However, he was not aware of the fact that, in going from unbuilt solution containing 2 grams of soap per liter to unbuilt solution containing 20 grams of soap per liter, there is an enormous increase in "colloidality" without the concomitant decrease in interfacial tension that would be predicted by his postulation of the part played by soap colloid. His data refute rather than confirm the interfacial activity of micelles.

In the previous discussion of the effect of alkaline builders on the surface tension of soap solutions, it was pointed out that the observed increase in tension is the net result of two opposing effects, namely, the effect of the hydroxyl-ion concentration to increase the surface tension and the effect of the total builder concentration to decrease

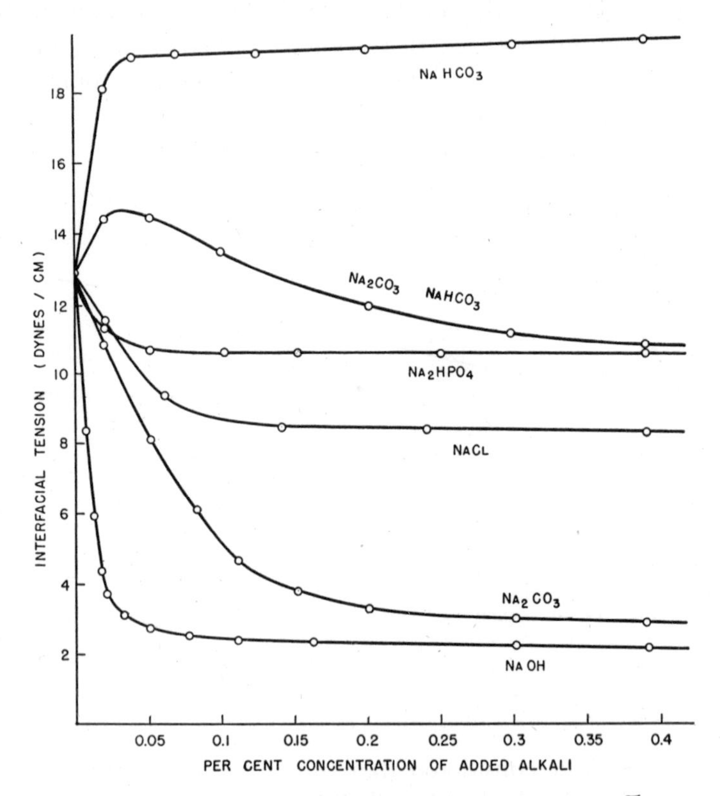

Figure 8–12. Influence of added salts on the interfacial tension of 0.1% sodium oleate solutions.[17]

the surface tension. In the case of interfacial tension, a somewhat similar condition probably exists, wherein the observed interfacial tension is the net result of two rather independent effects. By referring to Figure 8–12, from Powney and Addison,[17] it may be seen that, generally, builders which raise the pH of soap solutions lower

[17] Powney, J., and Addison, C. C., *Trans. Faraday Soc.*, **34**, 356–63 (1938).

the interfacial tension and that builders which lower the pH raise the interfacial tension. In other words, interfacial tension in such cases is somewhat of an inverse function of hydroxyl-ion concentration. However, this is by no means the complete picture. The definite lowering of interfacial tension by neutral sodium chloride (although less than that by salts giving high alkalinity), the leveling off of the increase in interfacial tension brought about by increasing concentrations of sodium bicarbonate, and the existence of a distinct maximum in the interfacial tension-modified soda curve, are all indicative of a tendency for the total concentration to lower the interfacial tension independently of the hydroxyl ions. This is a condition

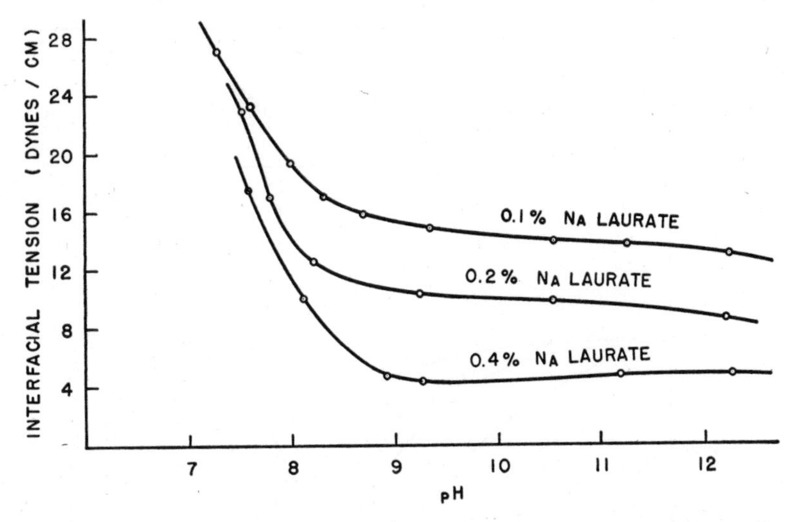

Figure 8–13. Interfacial tension-pH curves for sodium laurate solutions against xylene.[17]

analogous to that for surface tension. It is significant in that the lowering of interfacial tensions observed from the addition of alkaline builders and the increase observed from the addition of some acid salts are not entirely a function of hydroxyl-ion concentration. Rather, they are the net result of the independent effects of hydroxyl-ion concentration and total salt concentration.

Powney's and Addison's[17] results for the interfacial tensions of solutions of sodium laurate, potassium myristate, and sodium oleate against xylene, when buffered to various pH values by phosphate buffers, are shown in Figures 8–13 to 8–15. Their data do not indicate the actual buffer concentrations at the various experimental points

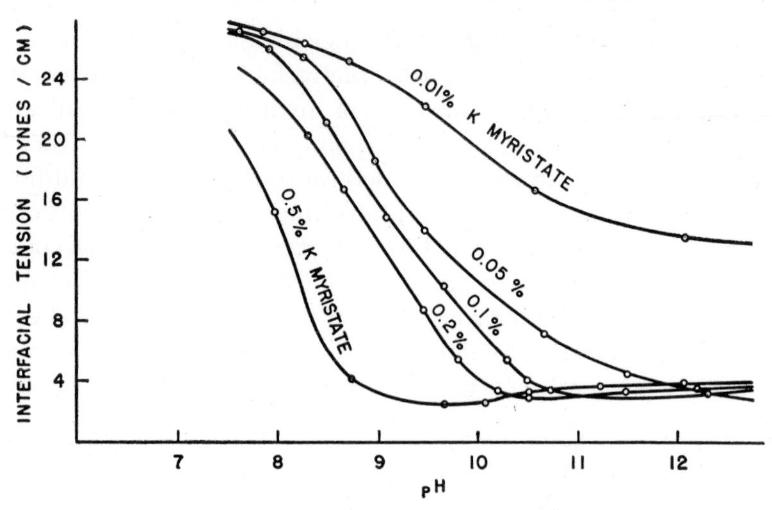

Figure 8–14. Interfacial tension-pH curves for potassium myristate solutions against xylene.[17]

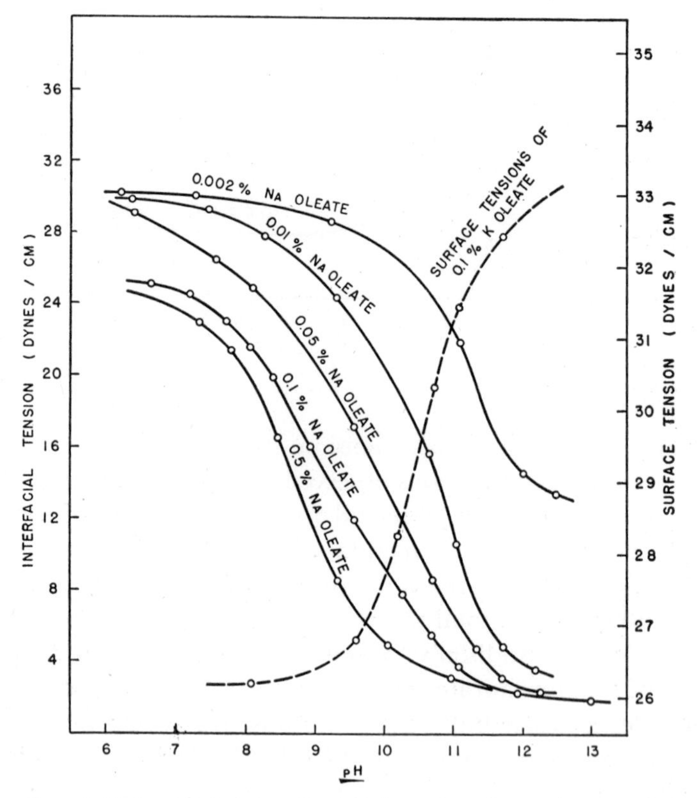

Figure 8–15. Interfacial tension-pH curves for sodium oleate solutions against xylene.[17]

which, as pointed out above, determine to some extent the observed interfacial tensions. However, on the premise that the hydroxyl-ion concentrations probably were much more significant than the effects of total salt concentration in these particular tests, the curves are approximately indicative of the effects of pH on the interfacial tensions of the solutions at the particular soap concentrations.

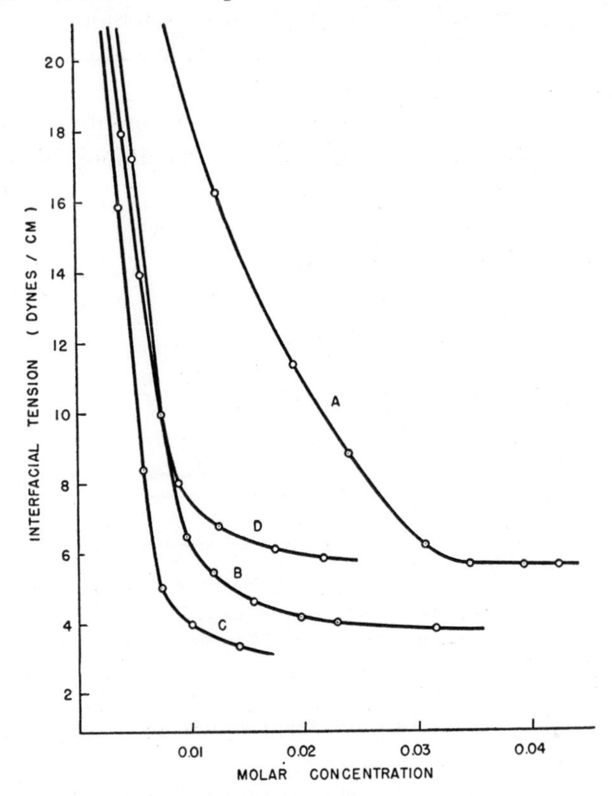

Figure 8–16. Interfacial tension-concentration curves for solutions of saturated soaps against xylene in the absence of builder (at 70°C).[18] Curves A, B, C and D are for sodium laurate, myristate, palmitate and stearate, respectively.

The effect of hydrolysis on the interfacial tensions of unbuilt soap solutions is illustrated in Figure 8–16.[18] Sodium laurate, with lesser hydrolysis in comparison to sodium myristate, palmitate, and stearate, gives an interfacial tension *vs.* concentration curve quite similar in form to that of an unhydrolyzed anionic synthetic detergent.

---

[18] Powney, J., and Addison, C. C., *Trans. Faraday Soc.*, **34**, 372–7 (1938).

This similarity is lost in the case of the soaps of higher molecular weight, particularly with the very highly hydrolyzed sodium stearate. If, now, alkaline builder is added to solutions of these soaps, the solutions all give interfacial tension-concentration curves quite similar in form to those of unhydrolyzed anionic detergents. This condition is illustrated by Powney's and Addision's[17] data shown in Figure 8–17 and must be indicative that suppression of hydrolysis is related to the manner in which alkaline builders affect the interfacial tension characteristics of soap solutions.

Data by Harris[7,10] which illustrate the pronounced lowering of the interfacial tension of solutions of several anionic detergents by

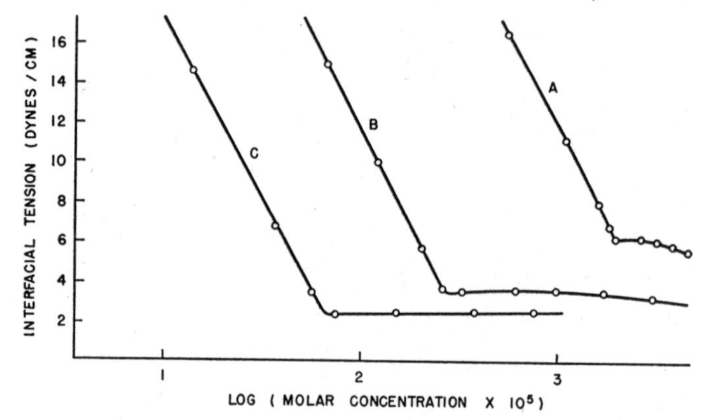

Figure 8–17. Interfacial tension-concentration curves for solutions of saturated soaps against xylene in the presence of 0.1% sodium hydroxide (at 70°C).[18] Curves A, B and C are for sodium laurate, myristate and palmitate, respectively.

various electrolytes already have been shown in Tables 8–1 and 8–2 (pages 145 and 146).

The depression of the interfacial tension of 0.005 per cent solutions of "Igepon T" ($C_{17}H_{33}CONCH_3CH_2CH_2SO_3Na$) by electrolytes of different cation valence has been studied by Robinson.[8] The concentrations of salt required to diminish the interfacial tension from 6.25 to 1 dyne per cm. were found to be .032*N* for NaCl, 0.003*N* for $CaCl_2$, and 0.00005*N* for $LaCl_3$, and are independent of the nature of the cation of a particular valence group.

Cassie and Palmer[19] have developed a mathematical treatment to account for the observed effects of builder cations of different

[19] Cassie, A. B. D., and Palmer, R. C., *Trans. Faraday Soc.*, 37, 156–68 (1941).

valences on the surface characteristics of solutions of strongly electrolytic anionic detergents. Their treatment, shown in part in the Appendix, is based on the ability of the builder cations to alter the potential energy of the fatty anions adsorbed in a surface or interface and, thus, to alter the extent of adsorption of these anions from the solution.

The total change in the energy of the system upon adsorption of a fatty anion in the surface is a function of the change in potential energy and of the change in electrical energy. Cassie and Palmer show that, for electrical effects to have a significant influence on the extent of adsorption, the product $z_i \epsilon \phi$ must be of the same order of magnitude as the change in potential energy, where z_i is the valence of ions whose presence exerts the electrical effect, ϵ is the charge on one electron, and ϕ is the electrostatic potential at the interface. They show, further, that the change of potential energy when one fatty anion of a chain length of C_{12} to C_{18} is adsorbed in the surface is of the order of $12kT$, where k is the Boltzmann constant (gas constant per ion) and T is the absolute temperature. The product $\epsilon\phi$ must then be of the order of $5kT$ for electrical effects to be significant.

An equation expressing the charge, σ, per unit area at the surface in relation to ions in the bulk of the solution, has been developed (see the Appendix) by Cassie and Palmer as

$$\sigma = \frac{DkT}{2\pi} \sum n_i (e^{-z_i \epsilon \phi / kT} - 1), \tag{8}$$

where D = dielectric constant for water and

n_i = the number of ions of valence z_i, per cc.

In the present case of adsorption of anions, both the charge per unit area, σ, and the potential, ϕ, are negative. Then the product $(-z_i \epsilon \phi)$ will be positive for cations and negative for anions. For univalent cations, the expression $e^{-z_i \epsilon \phi / kT}$ of equation (8) becomes $e^{(\epsilon\phi/kT)}$ and, if $\epsilon\phi$ must be of the order of $5kT$, the expression $e^{(\epsilon\phi/kT)}$ is of the order of 100 or more. Similarly, the effect of anions on the right-hand side of equation (8) is less than 0.01 and is negligible. In other words, only those ions of a sign opposite to that of the adsorbed ions have an appreciable effect on the charge at the surface. Cassie and Palmer have extended their derivation to obtain close quantitative agreement with observed experimental data.

Summarization of the most significant effects of builders on the interfacial tension of detergent solutions may be made as follows:

(a) Both alkalies and alkaline or neutral electrolyte salts may markedly lower the interfacial tensions of solutions of anionic detergents against oil.

(b) In the case of soap solutions, the effect of both alkalies and alkaline salts must be associated with suppression of hydrolysis. Presumably by this suppression, the curves for interfacial tension *vs.* concentration of soap assume the typical shape, together with sharp break at the critical concentration, of similar curves for unhydrolyzed anionic detergents.

(c) Builders which raise the pH of soap solutions tend to decrease the interfacial tension of such solutions against oil, and builders which lower the pH of the soap solutions generally increase the interfacial tension. The observed tensions are the net results of the independent actions of hydroxyl-ion concentration and of total salt concentration.

(d) The valence of the cation of electrolyte builders has a pronounced effect on the extent of lowering of the interfacial tension of solutions of anionic detergents; the ratio of cation concentration necessary to effect a given lowering may vary as much as 1 : 0.01 : 0.0001 when changing from univalent to bivalent to trivalent cations.

(e) It is both possible and practical to build detergent solutions to the extent where the proportionate amount of builder may even be predominant and yet obtain interfacial tensions in many cases lower than those of unbuilt solutions of equal total solute concentration. Unlike the case with surface-tension lowering, this possibility applies to both *soaps* and synthetic anionic detergents.

Builders and Micelle Formation

The foregoing discussions frequently refer to the effect of various builders on the critical micelle concentration for solutions of soaps and synthetic detergents, that is, on the concentration where micelle formation first becomes evident. The importance of this critical concentration in detergent processes justifies at this point a more extensive discussion of these effects.

It has been intimated already that the general effect of electrolyte builders on the critical micelle concentration is a marked reduction in the detergent concentration necessary to attain evident micelle formation. This effect has been studied by several investigators,[20-24] employing various means for detecting the critical points. Corrin and Harkins,[20] Klevens,[22] and Merrill and Getty[23] have observed that the addition of electrolyte salts of *univalent* cation to solutions of anionic detergents effects reductions in the critical concentration, the magnitudes of which depend only on the equivalent concentration of the added salt, independently of the nature of the anion of the salt. Thus, it is the concentration of the ion of the salt bearing a charge opposite to that of the micelle which determines the depression of the critical

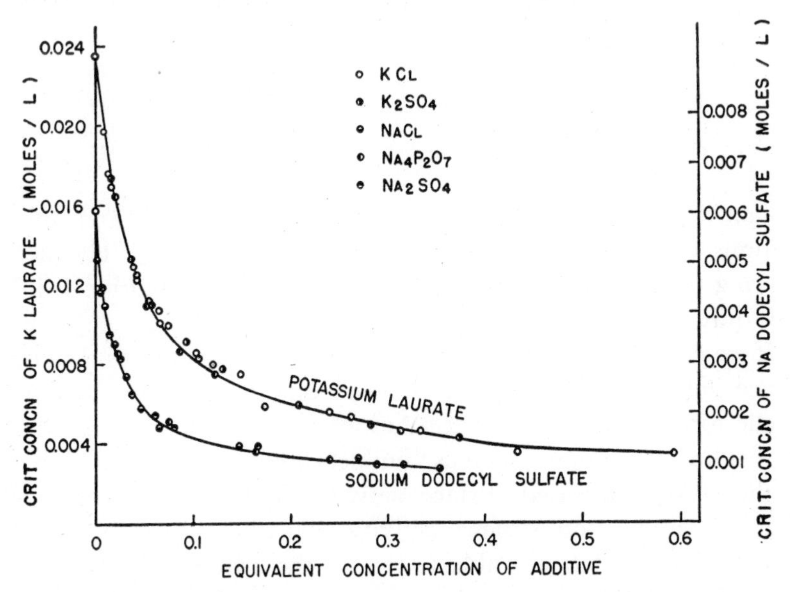

Figure 8–18. Effect of salts on the critical concentration for micelle formation in solutions of anionic detergents.[20]

concentration. This effect is the same whether the builder is an alkaline or a neutral salt, indicating that hydroxyl ion plays no significant part. These various effects are illustrated in Figure 8–18. Corrin and Harkins show further that, within a given series of de-

---

[20] Corrin, M. L., and Harkins, W. D., *J. Am. Chem. Soc.*, **69**, 683–8 (1947).

[21] Hartley, G. S., "Aqueous Solutions of Paraffin-Chain Salts," Paris, Hermann & Cie, 1936.

[22] Klevens, H. B., *J. Phys. & Colloid Chem.*, **52**, 130–48 (1948).

[23] Merrill, R. C., and Getty, R., *J. Phys. & Colloid Chem.*, **52**, 774–87 (1948).

[24] Wright, K. A., Abbott, A. D., Sivertz, V., and Tartar, H. V., *J. Am. Chem. Soc.*, **61**, 549–54 (1939).

tergents, the lower the unbuilt critical concentration (the greater the tendency to aggregate) the greater is the lowering of the critical concentration by equal amounts of salt.

Wright and coworkers,[24] in a study of the effects of sodium chloride on the critical concentration of sodium alkyl sulfonates, report that the per cent of lowering of the critical concentration becomes less marked with rise in temperature. However, examination of their data, as shown in Table 8–3, indicates that this effect is only due to

TABLE 8–3. EFFECT OF ADDED SODIUM CHLORIDE ON THE CRITICAL CONCENTRATION OF EQUIMOLAR SODIUM DODECYL SULFONATE SOLUTIONS [24]

Temp., °C	Critical Concentration Before, N_v	After, N_v	Difference, N_v
40.00	0.0110	0.0081	0.0029
60.00	0.0120	0.0092	0.0028
80.00	0.0146	0.0117	0.0029

the fact that the critical concentration of both the built and unbuilt solutions becomes higher with increasing temperature. The actual lowering occasioned by the addition of the sodium chloride is in all cases quite independent of the temperature.

If we consider micelles simply as a result rather than as a cause of surface activity, the effects of builders on surface activity and on micelle formation in detergent solutions must be closely related. An indirect line of reasoning such as "builders favor micelle formation and micelles favor increased surface activity" fails to account for the effect of builders on surface activity. Therefore, we should now examine *how* builders are able to reduce the critical micelle concentration and determine to what extent that "how" applies to the effects of builders on surface activity.

By way of review, the phenomenon of micelle formation in an anionic detergent solution is a manifestation that high free energies exist in the solution at the interface between solute and water and that, as one way to spontaneously reduce these free energies, the solute reduces its surface by aggregation. A certain "threshold" value of total free energy must be attained before such spontaneous aggregation starts. This threshold value is determined by the electrostatic repulsion between fatty anions. Such repulsion prevents aggregation until the forces driving the system towards aggregation exceed the forces of repulsion. The threshold total free energy in an unbuilt detergent solution must be that value prevailing just at the critical micelle concentration. In those solutions where the solute concentration is below the critical value, the summation of the surface areas of unaggregated solute and, thus, the total free surface energy are not sufficient to prompt spontaneous aggregation.

Lowering of the critical concentration necessary for the formation of micelles, as effected by electrolyte builders, conceivably might arise from a consequent increase in the free surface energy of the solute, from a reduction in the repulsion between solute "particles," or from a combination of the two. It is difficult to visualize that the effect of builders on the free energy of the solute can be more than an influence on the physical state in which the solute exists, with one physical state having a different free surface energy than another. Thus, we know that dissociation and hydrolysis of soaps are suppressed by electrolyte builders with the result that more of the soap either is present in the molecular state or is present as fatty ions of lowered activity. If, as is likely true, the free surface energy necessary to cause spontaneous aggregation to micelles is less for molecules than for fatty anions and is less for fatty anions the lower their activity, it then follows that the critical concentration for micelle formation will be reduced by any influence which favors the molecular state or causes suppression of the activity of the fatty anions. Herein lies probably the principal role of builders in the lowering of critical micelle concentration. The reduction of the activity of fatty anions by electrolyte builders, as discussed on page 137, may be likened to "neutralization" of the electrostatic repulsion between fatty anions.

We now must seek a possible parallelism between the role of electrolyte builders in aiding micelle formation and the role of builders in determining the surface activity of detergent solutions. Robinson[8] considers that, in solutions of anionic detergents, cations reduce the repulsion between heads of fatty anions adsorbed at an interface or surface, thus allowing a closer packing of the interfacial film and consequent decrease in interfacial or surface tension. Further, cations should reduce the potential barrier (negatively charged) existing at interfaces of anionic detergent solutions, thus enhancing entrance into the interface by similarly charged fatty anions. The conditions thus postulated by Robinson at interfaces are quite analogous to the conditions postulated above for micelle formation.

Pursuing the parallelism farther, we have already shown that builders can affect, under any conditions, the *degree* to which a particular detergent species is present in solution but that they can effect a complete *change in the identities* of the species only under certain limited conditions. The changes in surface activity effected by builders, as with change in concentration of builder, are not such as to indicate *changes in state of detergent*, except at or around the point of micelle formation. Rather, the changes in surface activity are such as to indicate changes in degree. Therefore, it is doubtful that a role can be established for builders, based on any change they can effect in the *identities* of the detergent species.

Robinson's postulation appears to account quite adequately for the effects of builders in reducing both the surface tension and interfacial tension of *unhydrolyzed* anionic detergents, and in reducing the interfacial tension of soap solutions. It also appears to account for the "mass" effect of total builder concentration which opposes the tendency of hydroxyl ions (from alkaline builders) to increase the surface tension of soap solutions. However, the converse argument that hydroxyl ions increase repulsion between adsorbed fatty anions and thus cause increased surface tension of soap solutions does not hold. Such an action, if it existed, also would take place in solutions of unhydrolyzed detergent. Therefore, hydrolysis products must play some significant part, as yet not definitely established, in the surface tension of soap solutions.

9

Influence of Builders on the Actions of Detergent Solutions

Through their influences on the nature and surface activity of detergents in solution, builders effect specific changes in the various actions of detergent solutions discussed in Chapter 5. These changes are of direct and practical importance to the operation of any detergent process.

Builders and Wetting

The wetting and spreading properties of aqueous fatty acid-sodium hydroxide solutions of different mol ratios have been studied extensively by Cupples,[1,2] whose results are shown in Figure 9–1, taken from his papers. The spreading coefficients as used in this figure are obtained by subtracting the sum of the surface tension plus interfacial tension from the surface tension of the reference mineral oil. For spontaneous wetting and spreading to occur it is necessary that the spreading coefficient, thus defined, be positive.

It is of particular interest to note from the surface-tension data of Figure 9–1 that not only does the surface tension increase substantially through a range above a mol ratio of 1.0, but also the minimum surface tensions for these solutions are all in the range where there is a considerable excess of fatty acid. In other words, not only does the addition of alkali to neutral soap solution raise the surface tension over a definite range, as discussed in Chapter 8, but the opposite

[1] Cupples, H. L., *Ind. Eng. Chem.*, **27**, 1219–22 (1935).

[2] Cupples, H. L., *Ind. Eng. Chem.*, **29**, 924–6 (1937).

condition of excess fatty acid produces the converse effect of lowered surface tension.

It is to be further noted from Figure 9–1 that in all cases the

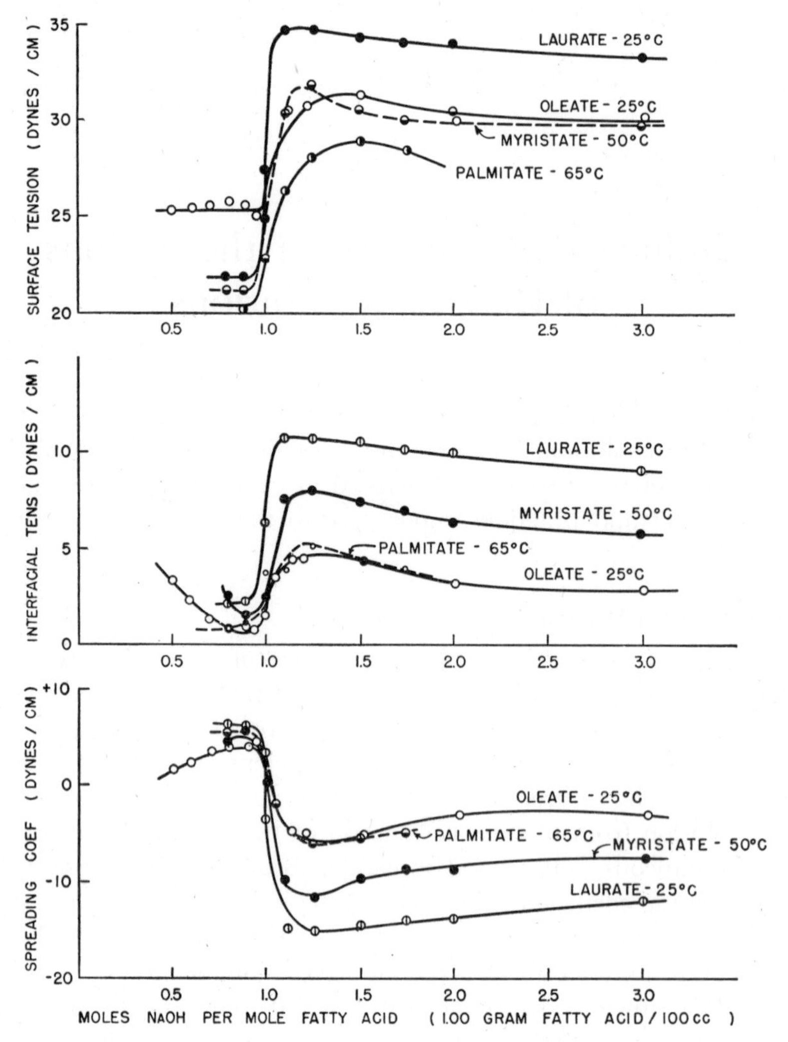

Figure 9–1. Variations in surface tension, interfacial tension against mineral oil, and spreading coefficient on mineral oil, with variations in sodium hydroxide-fatty acid ratio.[2]

interfacial tensions of the various series of solutions increase slightly over a narrow range above a mol ratio of 1.0. This apparent anomaly from the results discussed in Chapter 8 may well be due to difference in the nature of the oil phase. Thus, Cupples used medicinal petroleum oil and made the interfacial-tension measurements 10 minutes after formation of the interface. A slower rate of attainment of equilibrium at the interface when mineral oil is the opposite phase, as compared to the rate when liquids such as benzene are used, might well account for this variation from the results reported by Shorter,[3] Powney and Addison,[4] and others. In general, Cupples' data appear to substantiate the theory of Powney and Addison as to the difference in the nature of the significant surface-active species for surface-tension lowering and in that for interfacial-tension lowering as previously discussed in Chapter 4.

The net combined effect of sodium hydroxide on surface tension and on interfacial tension of soap solutions, as represented by the curves in Figure 9–1 for spreading coefficient, is a pronounced lowering in the tendency to wet and spread in the presence of excess alkali. For all the fatty acids above and including lauric acid, the maximum spreading coefficient is realized at a mol ratio of 1.0. Above this ratio the spreading coefficient decreases almost immediately to negative values. These results conform to those obtained by Powney[5] from determinations of the rates of penetration of sodium oleate solutions through cotton fabric in the presence of both alkali and neutral salt. Thus, by the use of sodium sesquicarbonate ($Na_4H_2(CO_3)_3$) so that the pH of the soap solution was unchanged (by buffering), the penetration rate for a solution of 0.1 per cent sodium oleate and 0.1 per cent alkali was practically the same as that of water alone. Addition of sodium chloride to the oleate solutions also markedly reduced the rate of penetration.

Powney and Frost[6] used contact angles at solution-solid interfaces as the indices of adhesional wetting and studied the influence of hydrogen-ion concentration on this property of soap solutions. Against paraffin wax they observed the results shown in Figure 9–2 (θ = contact angle) for solutions of sodium oleate of different concentrations, using 0.25 per cent phosphate buffers. It will be noted that the 0.01 per cent solution showed a minimum contact angle of

---

[3] Shorter, S. A., *J. Soc. Dyers Colourists*, **32**, 99–108 (1916).

[4] Powney, J., and Addison, C. C., *Trans. Faraday Soc.*, **34**, 356–63 (1938).

[5] Powney, J., "Wetting and Detergency Symposium," pp. 185–96, Brit. Sect. Intern. Soc. Leather Trades' Chem., 1937.

[6] Powney, J., and Frost, H. F., *J. Textile Inst.*, **28**, T237–54 (1937).

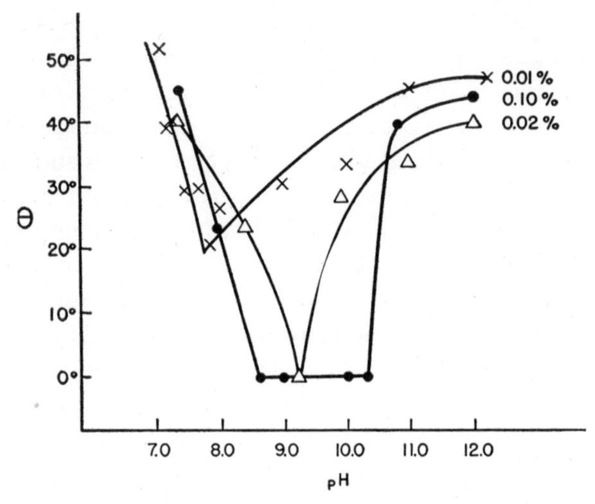

Figure 9–2. Contact angles at paraffin wax-air-sodium oleate solution interfaces, as functions of pH.[6]

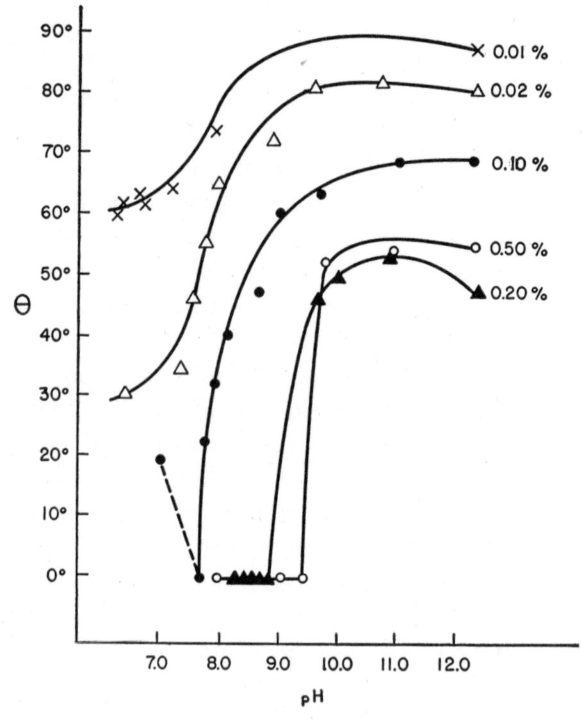

Figure 9–3. Contact angles at paraffin wax-air-sodium laurate solution interfaces, as functions of pH.[6]

about 20° at a pH slightly under 8.0 and that the 0.02 per cent solution showed a zero contact angle at a pH just above 9.0. Of particular significance is the fact that in each of these cases the pH range for minimum contact angle is very small. On the other hand, the 0.10 per cent sodium oleate solution showed zero contact angle over the approximate pH range of 8.5 to 10.3. Although not shown in the figure, the authors also report that a 0.2 per cent oleate solution gave a zero contact angle up to pH 12.25. Results quite comparable to those for sodium oleate solutions were reported by Powney and Frost for sodium laurate solutions at different concentrations and pH values. These results are shown in Figure 9–3. Comparison of the results in this figure with those in Figure 9–2 shows that, especially at low concentrations, the wetting by sodium laurate solutions is much inferior to that obtained with the same weight concentration of sodium oleate.

The above-discussed results of Powney and Frost permit the following generalizations of the effects of pH on the adhesional wetting of soap solutions:

(a) Going from a low pH to a high pH, the contact angle of wetting at first decreases, reaches a minimum, and then increases. In the regions of decreasing and increasing angles, the curve for contact angle *vs.* pH is generally steep.

(b) If the soap concentration is sufficiently high the pH range through which spreading will occur (zero contact angle) is quite broad.

Dreger and coworkers[7] have studied the effects of various electrolytes on the wetting time (canvas disc method) of a 2.27×10^{-3} molal sodium tridecane-7-sulfate solution, with the results shown in Figure 9–4. The wetting time for the unbuilt sulfate solution was so slow as to be out of the range of the test. By the addition of electrolytes the time was reduced to as low as 5 seconds and the effect increases as the valence of the cation of the electrolyte increases. However, this marked effect of added electrolyte is not typical of all such alkyl sulfate solutions. The particular sulfate mentioned above represents those whose unbuilt solutions have very steep wetting time *vs.* concentration curves. On the other hand, the wetting time of solu-

[7] Dreger, E. E., Keim, G. I., Miles, G. D., Shedlovsky, L., and Ross, J., *Ind. Eng. Chem.*, **36**, 610–7 (1944).

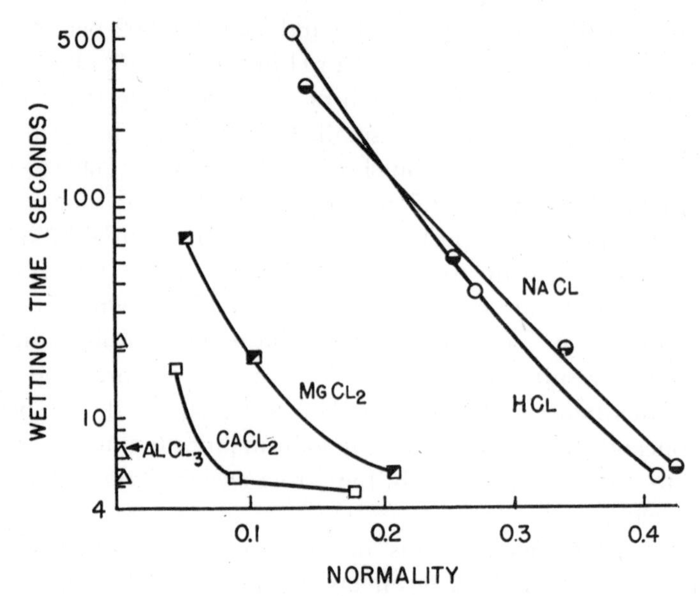

Figure 9–4. Canvas disc wetting time of 2.27 × 10^{-3} molal sodium tridecane-7-sulfate vs. concentration of added salt or acid, at 43.3–46.7°C.[7]

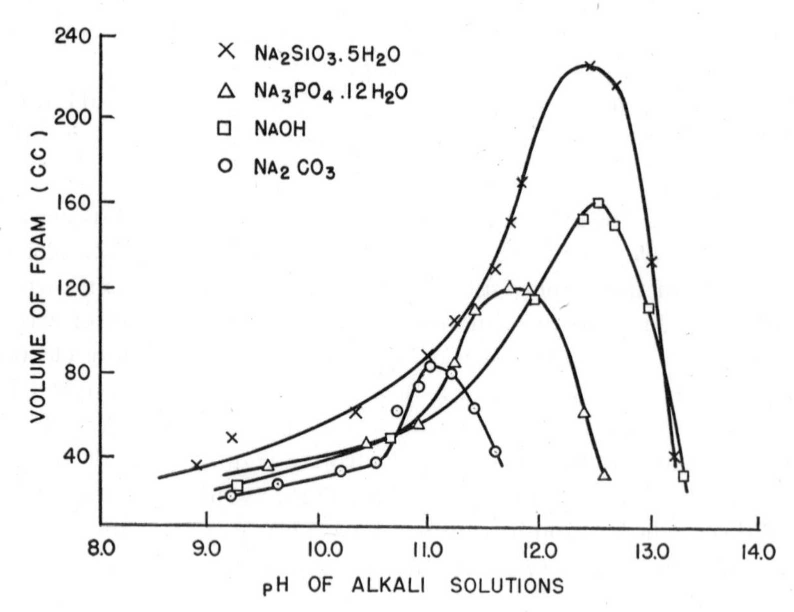

Figure 9–5. Volumes of foam formed on solutions of alkalies containing 0.0088% sodium stearate at 60°C, as functions of pH.[9]

tions of those alcohol sulfates which have relatively flat wetting time *vs.* concentration curves is very little affected by the addition of electrolyte.

Builders and Foaming

Stericker[8] has studied the effects of sodium silicates on the foaming or suds-forming powers of soap solutions. In general, the sudsing of soap solutions in which oil is emulsified is substantially improved by the addition of such a silicate as $Na_2O \cdot 3.3SiO_2$. Further, the stability of the foam when heated was greatly enhanced by the presence of the silicate. If the concentration is not too great, sodium silicate stabilizes the suds formed by soap.

In an extensive comparison of several alkaline builders, Baker[9] has studied the foaming of solutions of different concentrations of alkalies containing 0.0088 per cent sodium stearate at 60°C. His results are shown in Figures 9–5 and 9–6, the first being plotted with foam volume *vs.* pH and the second being plotted with foam volume *vs.* weight per cent of alkali. As pointed out by Baker, the most striking feature of these curves is that they show decided optimum values. He presents an explanation of the increases to the maxima with increased pH or concentration of alkali, based on the greater suppression of hydrolysis of the soap and the greater conversion of soap to the colloidal form. He attributes the sharp declines in foam volume after the maxima to the salting-out effect of the relatively high concentrations of alkali in these ranges. Vail[10] has recalculated the data of Baker to a basis of sodium normality. As shown in Figure 9–7, the maxima for the different alkalies on this basis occur more nearly at the same concentrations.

In a recent study of the effects of pH on the foaming characteristics of 0.1 per cent soap solutions, Miles and Ross[11] report the data shown in Figure 9–8 for sodium caprate, laurate, myristate, and palmitate. The pronounced maxima reported by Baker are here confirmed, as well as the fact that with soaps of higher fatty acids the region of the maxima covers a wider pH range. Miles and Ross have extended this study by selecting from their pH data the pH for

---

[8] Stericker, W., *Ind. Eng. Chem.*, **15**, 244–8 (1923).
[9] Baker, C. L., *Ind. Eng. Chem.*, **23**, 1025–32 (1931).
[10] Vail, J. G., *Ind. Eng. Chem.*, **28**, 294–9 (1936).
[11] Miles, G. D., and Ross, J., *J. Phys. Chem.*, **48**, 280–90 (1944).

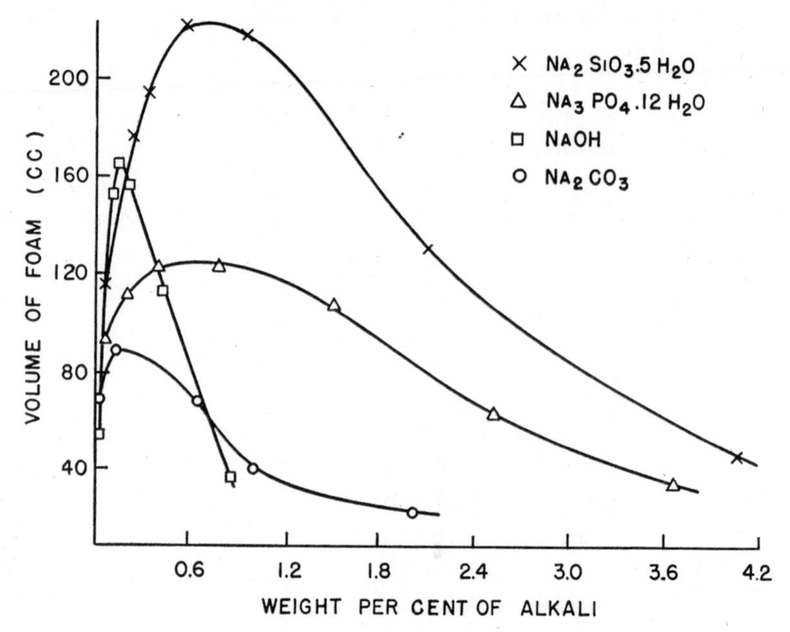

Figure 9–6. Volumes of foam formed on solutions of alkalies containing 0.0088% sodium stearate at 60°C, as functions of weight % of alkali.[9]

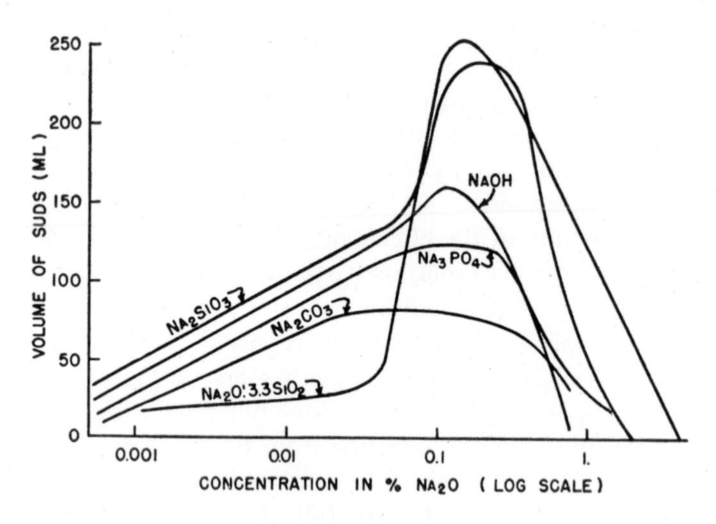

Figure 9–7. Volumes of suds formed on solutions of alkalies containing 0.0088% sodium stearate at 60°C. vs. units of alkali, to show anion effects (recalculated from Baker).[10]

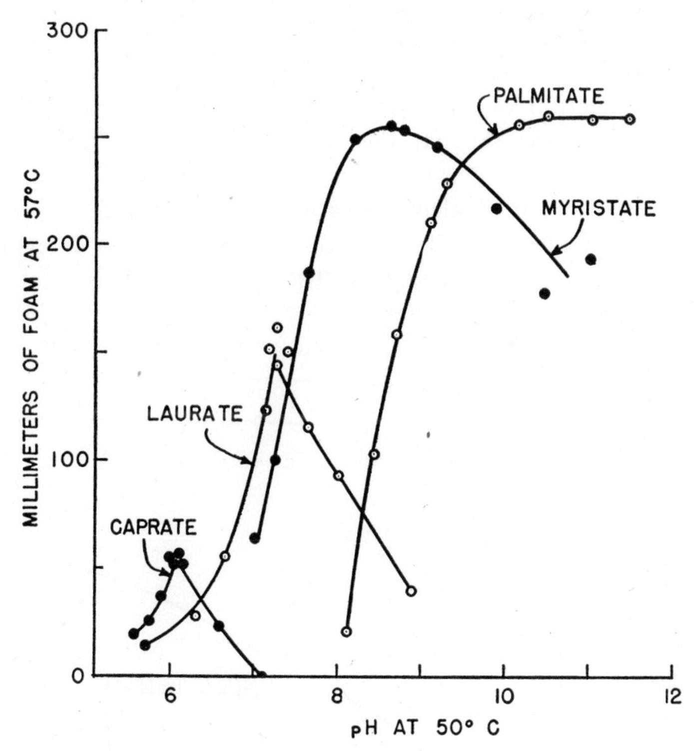

Figure 9–8. Depths of foam formed on 0.1% solutions of saturated sodium soaps at 57°C, as functions of pH.[11]

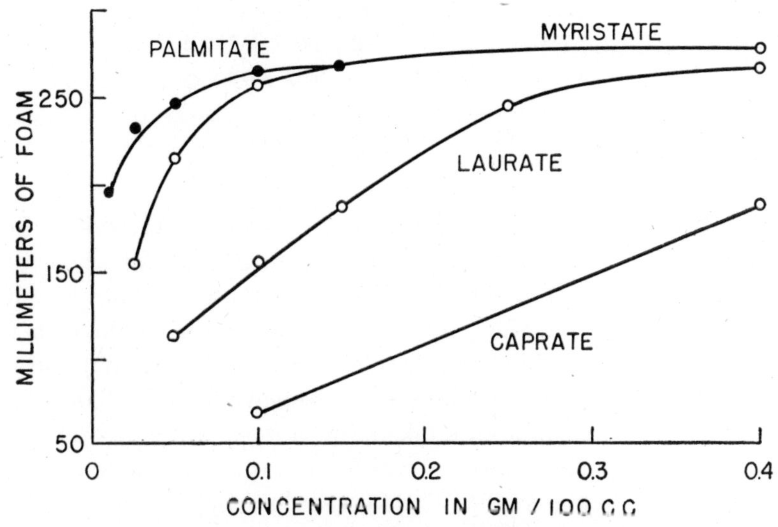

Figure 9–9. Depths of foam formed on solutions of *saturated* sodium soaps at 57°C, as functions of concentration when each solution is adjusted to the pH for maximum foam stability.[11]

maximum foam volume for each of the 0.1 per cent soap solutions and determining the foaming characteristics of solutions of those soaps as a function of soap concentration where each solution is adjusted to the optimum pH. The results of this study are shown in Figures 9–9 and 9–10.

The effects of added electrolytes on the foaming of dilute solutions of sodium alcohol sulfates have been studied by Dreger and coworkers.[7] Their results for 2.27×10^{-3} molal aqueous solutions of

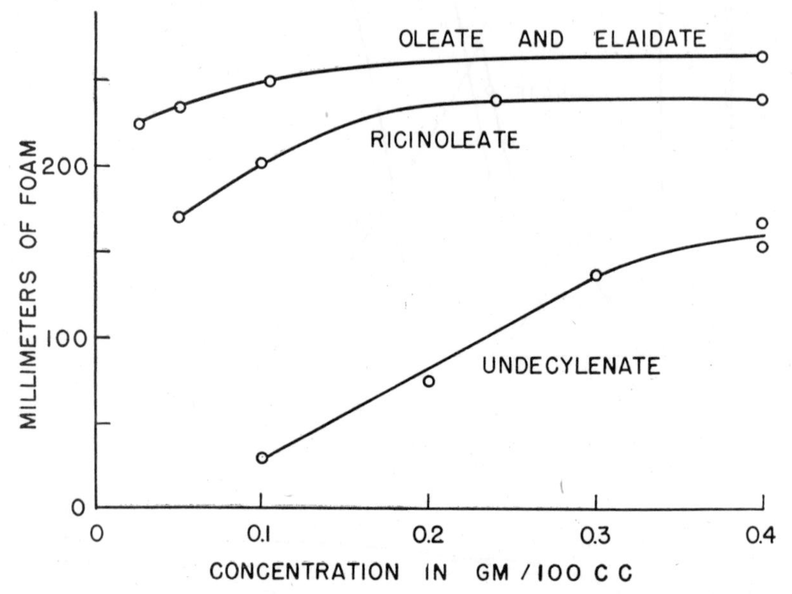

Figure 9–10. Depths of foam formed on solutions of *unsaturated* sodium soaps at 57°C, as functions of concentration when each solution is adjusted to the pH of maximum foam stability.[11]

sodium tridecane-7-sulfate are shown in Figure 9–11, taken from their data. It should be pointed out in connection with these data that the foams of all the solutions, except those containing $AlCl_3$ above 0.005*N* (in 0.0137*N* HCl), broke down to a height of a few millimeters within 5 minutes. Dreger found, however, that the salt effects lead to more stable foams when somewhat higher concentrations of the alcohol sulfates are taken, but not above the steep portion of the foam *vs.* concentration curves for the unbuilt sulfate solutions (see Figure 5–5, Chapter 5). The foam-enhancing effects of sodium chloride and of magnesium chloride in solutions of several alcohol

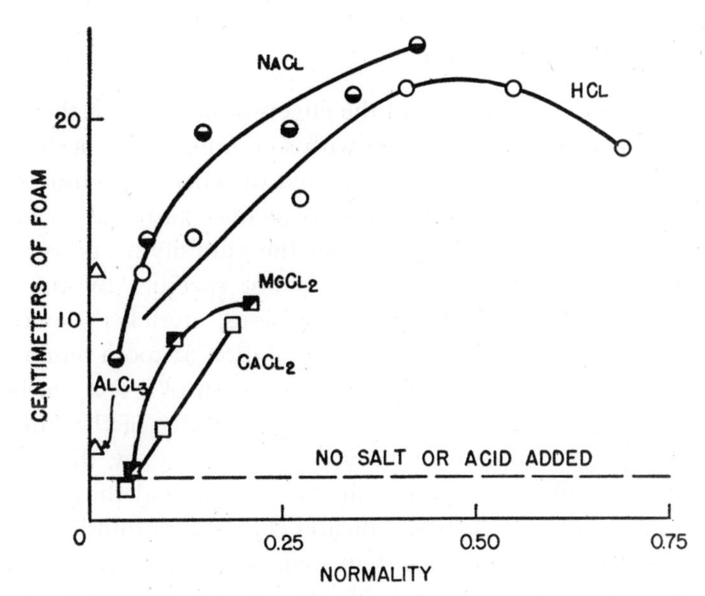

Figure 9–11. Depth of foam vs. normality of salt or acid added to 2.27×10^{-3} molal solutions of sodium tridecane-7-sulfate at 46°C.[7] (See text for stability of these foams.)

sulfates are shown in Table 9–1, as reported by Dreger. The pronounced increase in amount and stability of foam under these conditions is easily apparent.

TABLE 9–1. EFFECTS OF SODIUM CHLORIDE AND MAGNESIUM CHLORIDE ON FOAM OF ALCOHOL SULFATES[7]

Sodium Alcohol Sulfate	Concentration, Molal $\times 10^4$	Concentration of Salt: %	Concentration of Salt: Normality $\times 10^3$	Max. mm. of Foam at 115°F (46°C): Initial	Max. mm. of Foam at 115°F (46°C): After 15 min.	Ratio, NaCl : $MgCl_2$ for Maximum Effect on Foam: % basis	Ratio, NaCl : $MgCl_2$ for Maximum Effect on Foam: Normality basis
Dodecyl	20			165	10	33 : 1	27 : 1
sulfate		0.5 NaCl	86	220	205		
		.015 $MgCl_2$	3.15	220	205		
Tetradecyl	4			140	15	25 : 1	20 : 1
sulfate		0.025 NaCl	4.3	210	205		
		.001 $MgCl_2$	0.2	210	200		
Tridecane-2-	10			140	10*	33 : 1	27 : 1
sulfate		0.5 NaCl	86	210	185		
		.015 $MgCl_2$	3.15	200	170		
Pentadecane-	2.5			150	30	12.5 : 1	10 : 1
2-sulfate		0.025 NaCl	4.3	200	185		
		.002 $MgCl_2$	0.42	190	185		
Pentadecane-	20			200	10	50 : 2	42 : 1
8-sulfate		0.25 NaCl	43	245	245		
		.005 $MgCl_2$	1.05	235	225		

* After 5 minutes

Builders and Emulsification

Some of the earliest studies of the effects of builders on the emulsifying power of soap solutions were with sodium silicates as the builders. Thus, Richardson[12] in 1923 correlated the effect of sodium silicate on the drop number of soap solutions against kerosene as being a measure of the effect of the silicate on the emulsifying power of the solutions. However, this correlation is not specific. By the direct method of preparing oil-in-solution emulsions, Stericker[8] determined that neutral stove oil emulsions in soap solution at room temperature were not appreciably improved by the addition of a few tenths of a per cent of sodium silicate. However, the stability of the emulsions when heated was definitely improved, particularly by the presence of the less basic silicates. In general, Stericker found that a mixture of soap and one of the less basic silicates is a better emulsifying agent for mineral or saponifiable oils than soap alone or soap plus sodium carbonate. Particularly stable, finely dispersed emulsions are formed with silicate-soap mixtures when the oil contains acidic or saponifiable material.

Vincent[13] considers alkali salts to improve the emulsifying power of soap solutions because of the lowering in interfacial tension which they effect against oil. He found sodium hydroxide, trisodium phosphate, and ammonium hydroxide to be more efficient than sodium carbonate and borax and that a compound containing 20 per cent soap and 80 per cent sodium silicate (Na_2O/SiO_2 ratio about 3) had excellent emulsifying power. He also found that the emulsifying action of soap passed through a maximum as increasing amounts of sodium hydroxide and trisodium phosphate were added to the solution. He considered this reduction in emulsification after the optimum point to be due to adsorption of sodium ions. However, Rhodes and Bascom[14] show that this cannot be the case, from the fact that the hydroxide, the phosphate, and the carbonate all exhibit their maximum effects at the same alkalinity, although the concentrations of the three substances and of the sodium ions derived from them are widely different at this optimum point.

---

[12] Richardson, A. S., *Ind. Eng. Chem.*, **15**, 241–3 (1923).
[13] Vincent, G. P., *J. Phys. Chem.*, **31**, 1281–315 (1927).
[14] Rhodes, F. H., and Bascom, C. H., *Ind. Eng. Chem.*, **23**, 778–80 (1931).

In comparative tests between builders, Mitchell[15] found that, at concentrations of interest in detergent processes, mixtures of sodium oleate with a silicate form a larger volume of emulsion with paraffin oil than does the same soap with caustic soda, soda ash, sodium sesquicarbonate, trisodium phosphate, sodium aluminate, or borax. The latter, however, was equally effective at higher concentrations under his conditions.

An indirect but none the less significant influence of several builders, particularly the phosphates and silicates, on emulsification by soaps is that brought about by the sequestering or removing of Ca^{++} ions and Mg^{++} ions present as hardness in the water. By such action reversion of emulsions by polyvalent cations, as discussed in Chapter 6, may be averted, and loss of charge on the emulsion droplets by the neutralizing action of such cations can be prevented.

In connection with studies of the combined deflocculating (stabilizing) and emulsifying powers of builder solutions and of built soap solutions, Snell[16] has employed oil-coated solid particles on the premise that the oil-coated particle would act as an oil droplet in the dispersed state. This technique, to be truly representative of an emulsification process, requires that the solid particle remain within the oil film in the suspension and exert no influence on the properties of the exterior oil surface. It is doubtful that these requirements have been fully met in the various reported studies; however, the studies undoubtedly contribute much to our knowledge of the role of builders in detergent processes because of the close parallelism between the experimental conditions and the conditions actually encountered in, say, laundering.

Builders and Solubilization

Solubilization of water-insoluble materials in the micelles of detergent solutions can very well be expected to be markedly affected by the presence of builders because of the pronounced effect of the builders on micelle formation. Thus, Hartley[17] found that the maximum solubility ratio of trans-azobenzene in solutions of cetylpyridinium chloride (ratio of the solubility divided by the normality of the

---

[15] Mitchell, R. W., *Metal Cleaning Finishing*, **2**, 389–94 (1930).

[16] Snell, F. D., *Ind. Eng. Chem.*, **25**, 162–5 (1933).

[17] Hartley, G. S., "Aqueous Solutions of Paraffin-Chain Salts," Paris, Hermann & Cie, 1936.

cetylpyridinium chloride solution) could be reached at a lower concentration of the detergent in the presence of sodium chloride, as shown in Figure 9–12.[18]

In work by McBain and O'Connor[19] it was found that $Na_2O \cdot 3.19SiO_2$ and potassium hydroxide each increased the solubilizing power of potassium oleate solutions for the vapors of volatile hydrocarbons. On the other hand, the addition of these materials to the oleate solution reduced the solubility of the liquid-phase volatile hydrocarbon, as evidenced by the fact that the partial pressure of

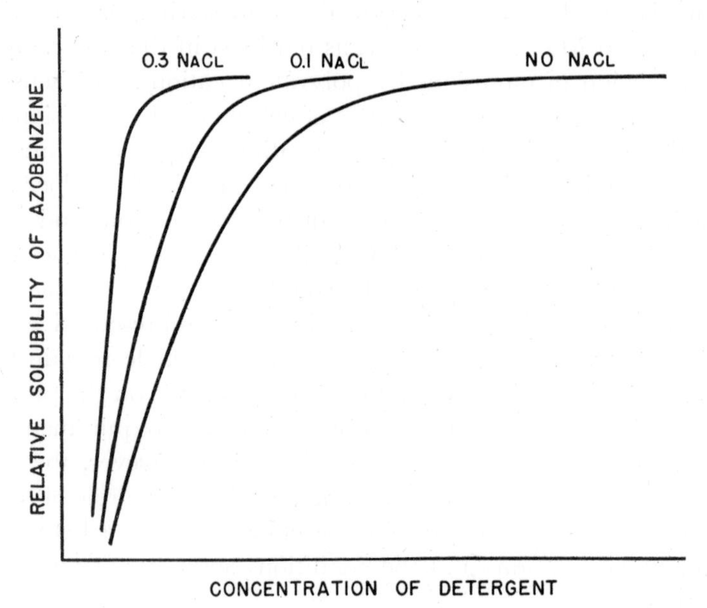

Figure 9–12. Effect of electrolytes on solubilization of trans-azobenzene in solutions of cetylpyridinium chloride.[18]

the hydrocarbon was greater than in the presence of potassium oleate alone. However, the soap solutions were of a concentration of 0.35 per cent and higher, so that a salting-out effect by the alkali was probably realized. McBain and Merrill[20] have pointed out that, in concentrations where the solubilizer is not fully colloidal, the addition of salts increases the solubilization action for water-insoluble dyes because of its effect of increasing the formation of micelles. On the other hand,

[18] Tomlinson, K., *Mfg. Chemist*, **15**, 198–200, 201 (1944).
[19] McBain, J. W., and O'Connor, J. J., *J. Am. Chem. Soc.*, **62**, 2855–9 (1940).
[20] McBain, J. W., and Merrill, R. C., Jr., *Ind. Eng. Chem.*, **34**, 915–9 (1942).

at concentrations where micelle formation is already optimal, the addition of salts may reduce solubilization. McBain and Johnson[21] consider that potassium chloride not only greatly increases the solubilizing power of fully formed laurate and myristate micelles but it produces in dilute solution micelles of still higher solubilizing power.

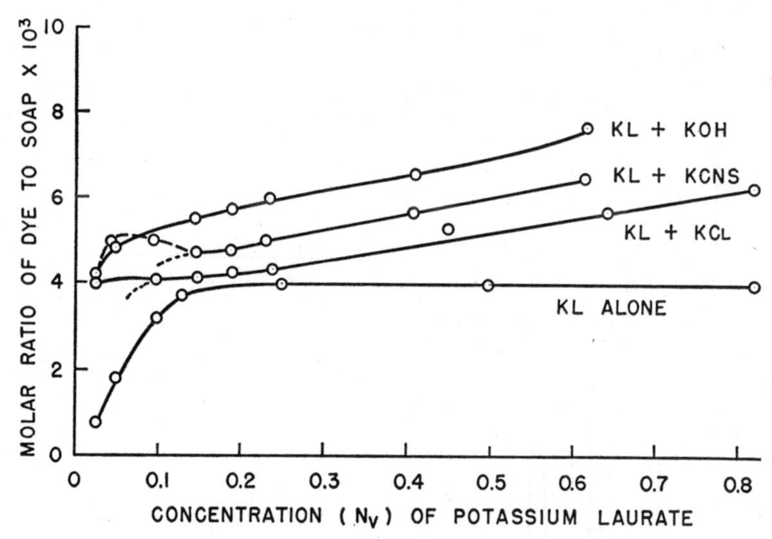

Figure 9–13. Effect of 1M potassium hydroxide, thiocyanate and chloride on solubilization of Orange OT dye by potassium laurate solutions at 25°C (concentration of KL expressed as N_V).[22]

Data by McBain and Green[22] for the effects of added salts on the solubilization of water-insoluble dye are shown in Figures 9–13 and 9–14, taken from their data.

Builders and Stabilization

The earliest direct studies of the stabilizing effect of alkaline builders on suspensions of solid soil in soap solution employed the technique of Spring,[23] namely, the power of the builders to aid the passage of the suspended material through filter paper. Thus, Elledge and Isherwood[24] in 1916 used suspensions of lampblack in 0.4 per cent soap solutions to which various alkalies were added in a con-

[21] McBain, J. W., and Johnson, K. E., *J. Am. Chem. Soc.*, **66**, 9–13 (1944).
[22] McBain, J. W., and Green, Sister A. A., *J. Am. Chem. Soc.*, **68**, 1731–6 (1946).
[23] Spring, W., *Rec. Trav. Chim.*, **28**, 120–35 (1909).
[24] Elledge, H. G., and Isherwood, J. J., *Ind. Eng. Chem.*, **8**, 793–4 (1916).

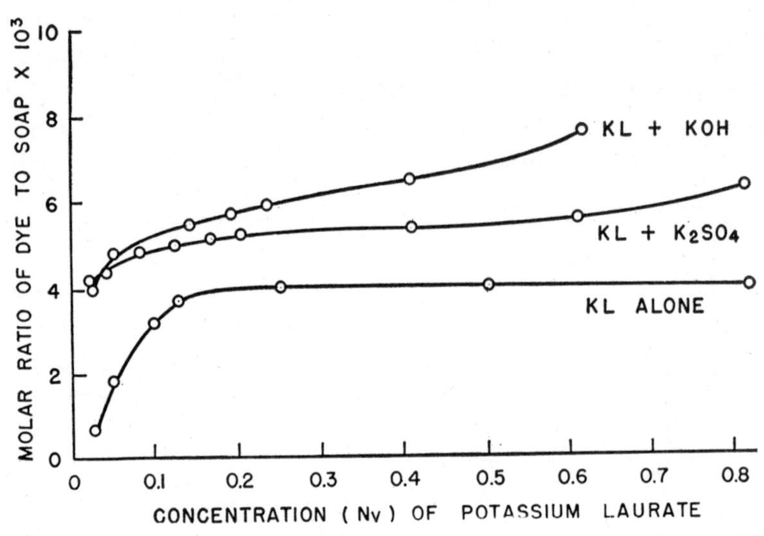

Figure 9–14. Effect of 1*M* potassium hydroxide and sulfate on solubilization of Orange OT dye by potassium laurate solutions at 25°C (concentration of KL expressed as N_V).[22]

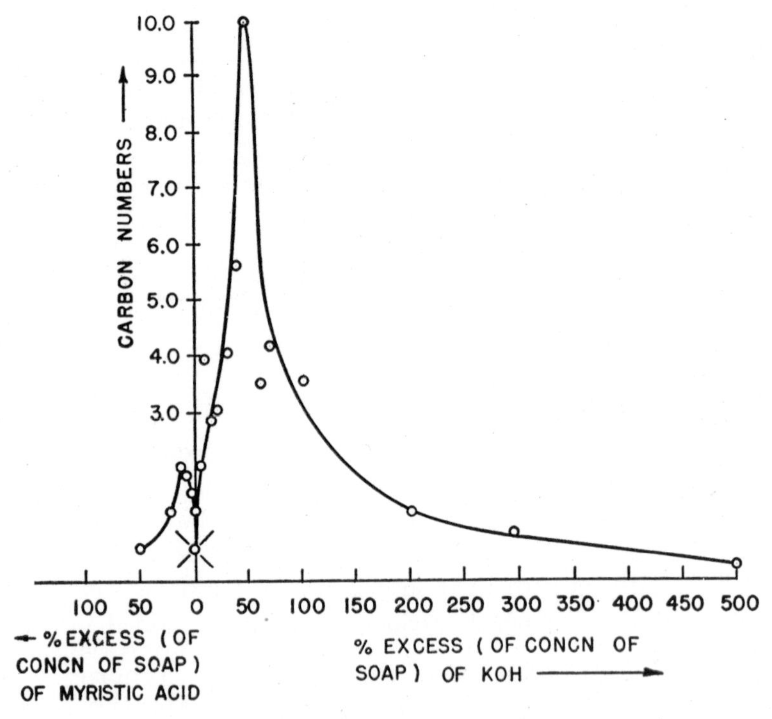

Figure 9–15. Variation of carbon number of 0.125N_W potassium myristate solutions, as a function of excess of potassium hydroxide or of myristic acid.[25]

centration equivalent to 0.5 per cent sodium carbonate. On passing their suspensions through filter papers, they found that the greatest amount of lampblack would pass through the filter paper in those cases where the lowest interfacial tension had been attained by the addition of alkali. They thus correlated stability with interfacial tension, a correlation which has since been found not strictly accurate. McBain and coworkers[25] used carbon black suspensions in similar studies and considered that the carbon black carried through the paper was a measure of that stably suspended. Their results with 0.125*N* solutions of potassium myristate to which excesses of potassium hydroxide or of myristic acid were added are shown in Figure 9–15. It will be noted that slight excesses of either the alkali or the fatty acid increase the stabilizing action many times, the excess alkali particularly so.

Fall,[26] in a study of suspensions of manganese dioxide, reports that silicates of sodium as addition agents to soap solutions whose unbuilt concentrations are below the value for optimum stabilization enhance the stabilizing powers of the soap more than other alkalies, such as NaOH, Na_2CO_3, and Na_3PO_4. He attributed this to the silicates being more nearly like soap than the other alkalies, particularly with respect to their physical form.

The extent of deposition of pigments from suspension onto cotton cloth has been used by Carter[27] as an index of the stability of the suspension. He prepared suspensions of several pigments in solutions whose total detergent content consisted of different proportions of sodium oleate and various alkaline builders. By immersing cotton test pieces in the suspensions and then determining the percentage retention of original whiteness, he obtained the results shown in Figures 9–16 and 9–17 with ferric oxide soil and carbon black soil, respectively. Recalculation of the detergent compositions used by Carter, whereby the proportion of alkaline builder is expressed as equivalent Na_2O, makes possible the plots shown in Figures 9–18 and 9–19. It is of particular interest to note that the stabilizing power of unbuilt soap solution for these soils in most cases is impeded by the substitution of alkaline builder for part of the soap, even though several of

[25] McBain, J. W., Harborne, R. S., and King, A. M., *J. Soc. Chem. Ind.*, **42,** 373–8T (1923).

[26] Fall, P. H., *J. Phys. Chem.*, **31,** 801–49 (1927).

[27] Carter, J. D., *Ind. Eng. Chem.*, **23,** 1389–95 (1931).

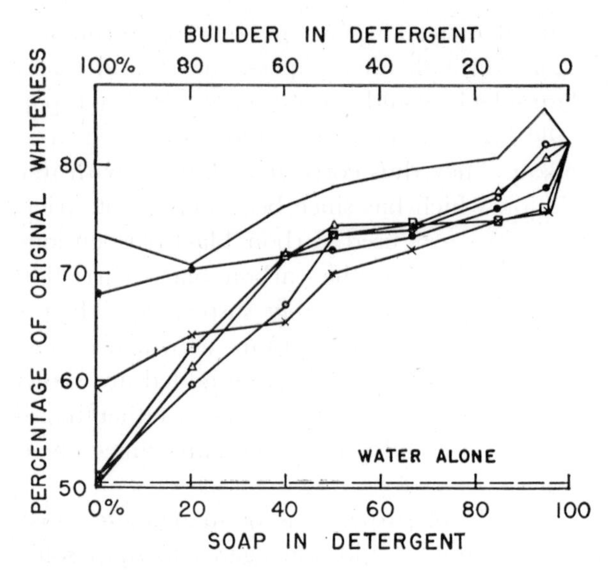

Figure 9–16. Deposition of ferric oxide on cotton cloth with various builder-sodium oleate combinations.[27]

●—$Na_2O \cdot 3.25SiO_2$;
⬤—Na_2SiO_3;
×—Na_3PO_4;
△—Na_2CO_3;
○—NaOH;
□—Modified soda ($NaHCO_3/Na_2CO_3$ = 2.23 by weight).

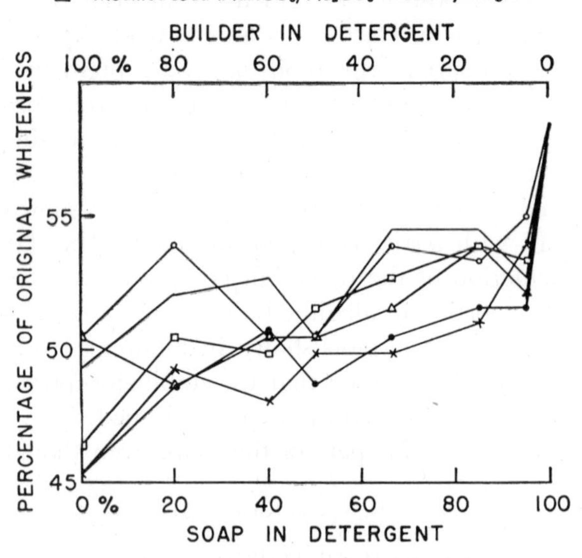

Figure 9–17. Deposition of carbon black on cotton cloth with various builder-sodium oleate combinations.[27] (For identity of various curves see Figure 9–16.)

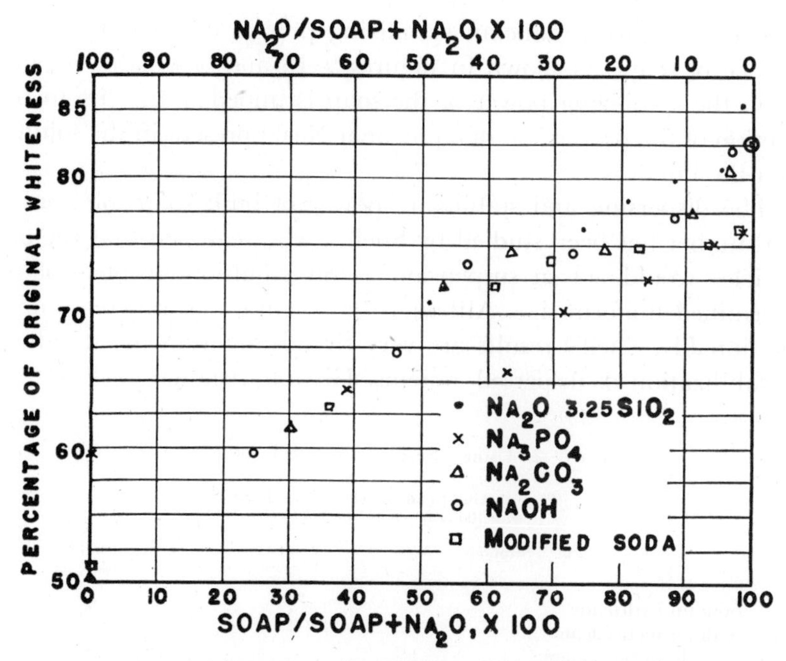

Figure 9–18. Data of Figure 9–16 recalculated to show relationship between deposition of ferric oxide and equivalent Na_2O content of detergent solution.

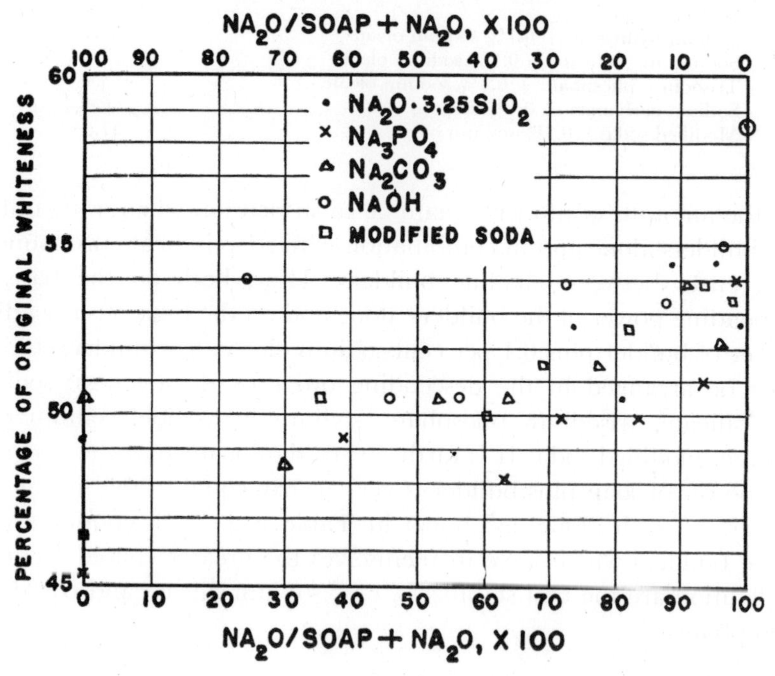

Figure 9–19. Data of Figure 9–17 recalculated to show relationship between deposition of carbon black and equivalent Na_2O content of detergent solution.

the builders have significant stabilizing powers of their own in the absence of soap. As shown in Figures 9–18 and 9–19, the extent to which the stabilizing power of the soap is impeded appears to be a function of the proportion of equivalent Na_2O present in the solutions from the builders.

The dispersing and stabilizing power of built soap solutions for oiled umber has been studied by Snell,[16] who reasoned that the oiled particles would act in suspension as oil emulsion droplets of predetermined uniform size. Although the accuracy of this reasoning is questionable, Snell's results are very significant from the standpoint of stabilization of dispersed systems. His data, shown in Table 9–2,

TABLE 9–2. UMBER SUSPENDED AFTER 24 HOURS [16]

(From shaking 1 gram of oiled umber at 20°C with 100 cc of emulsifying solution)

0.1% Solutions	Umber per 100 cc Suspension, mg
Sodium hydroxide	18.9
Sodium metasilicate	32.2
Trisodium phosphate	25.3
Sodium carbonate	12.2
Modified soda	12.2
Sodium oleate	48.6
Sodium hydroxide + 0.1% sodium oleate	66.4
Sodium metasilicate + 0.1% sodium oleate	77.0
Trisodium phosphate + 0.1% sodium oleate	73.4
Sodium carbonate + 0.1% sodium oleate	66.4
Modified soda + 0.1% sodium oleate	57.6

for the amounts of umber remaining in suspension when using different builders alone and in combination with soap indicate pronounced differences between alkaline builders. Thus, both on the basis of suspending power of the builders alone and on the basis of suspending power of builder plus 0.1 per cent sodium oleate, the builders studied may be arranged in the descending order of effectiveness: sodium metasilicate, trisodium phosphate, sodium hydroxide, sodium carbonate, modified soda. It is further very significant that the suspending power of soap plus builder is very nearly equal to the sum of the powers of each alone, as shown in Table 9–3. Snell considers that those builders which have in themselves the greater colloidal nature in solution are the best stabilizers, e.g., sodium silicate and trisodium phosphate.

Baker[9] has determined the amount of soap necessary to produce a stable suspension of bone black in 0.4 per cent solutions of sodium metasilicate in comparison to the amount required in suspensions of bone black in water. He found that decidedly less soap was necessary to effect stabilization in the presence of the silicate.

The nature of the stabilizing action of builders, alone or with soap, has been considered by McBain.[28] He suggests that the answer as to how coarse particles may be made to behave in suspension like fine particles without changing their actual size depends upon imparting a charge to the particle so that ions of opposite sign to

TABLE 9–3. COMPARISON OF SUSPENDING POWER OF ALKALINE SALT AND SOAP SOLUTIONS AT 20°C [16]

0.1% Solutions	Umber Suspended, mg./100 cc Sum of values for alkali alone, plus 48.6, value for sodium oleate alone	Value determined for alkali and sodium oleate together
Sodium hydroxide	67.5	66.4
Sodium metasilicate	80.8	77.0
Trisodium phosphate	73.9	73.4
Sodium carbonate	60.8	66.4
Modified soda	60.8	57.6

those on the particle must remain in the vicinity through electrostatic attraction. As McBain pictures the condition, ". . . when the particle tries to settle, it must carry along with it all its atmosphere of free ions. Then the effective weight of each discrete unit in the system, that is, particle plus each ion, is the average of all. For example, if there are n charges on a particle, say 100, and n corresponding free ions, the average weight would be approximately $1/(n + 1)$ or $\frac{1}{101}$ of that of the original uncharged particle, since the ions are so small in comparison with the particle that their individual weights can be neglected. The effect is similar to a stone falling in water, but attached to a large number of small floats not sufficient actually to buoy it up, but enough to keep it held up in eddy currents." McBain further postulates that soap can combine protective action with the suspending action of the electrolyte ions and, therefore, be much more effective.

However, it is doubtful that the foregoing discussion tells the complete story of stabilization by builders. The fact that the stabilizing

[28] McBain, J. W., "Solubilization and Other Factors in Detergent Action," "Advances in Colloid Science," pp. 99–142, New York, Interscience Publishers, Inc., 1942.

power of soap alone is sometimes enhanced by the addition of electrolyte to the system or even by substitution of electrolyte for part of the soap may very well be associated with such a factor as the effect of the electrolyte on the physical state of the soap. Further, the data of Powney and Wood[29] for the electrophoretic mobility of oil droplets

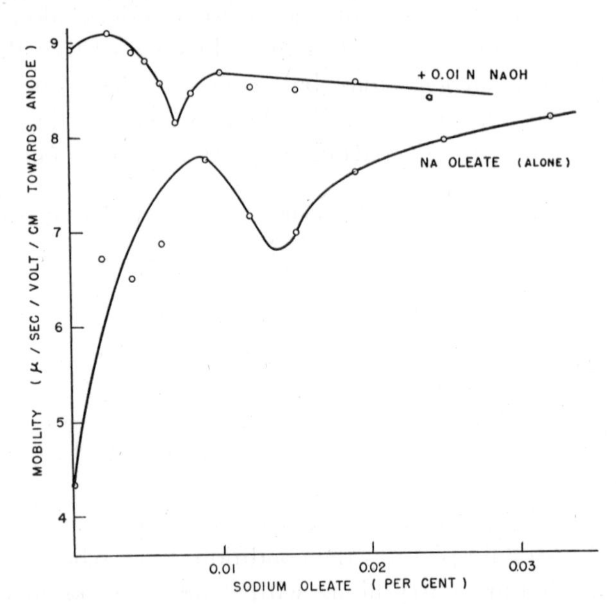

Figure 9–20. Mobility of dispersed oil droplets vs. concentration of sodium oleate, alone and built with sodium hydroxide.[29]

in soap solutions and the effects produced thereon by added alkali are of considerable interest. The increase in mobility effected by 0.01*N* sodium hydroxide added to sodium oleate solutions is apparent from Figure 9–20, taken from their data. The increased mobility may be considered a function of the increased zeta potential imparted to the oil droplets by the alkali.

---

[29] Powney, J., and Wood, L. J., *Trans. Faraday Soc.*, **36,** 57–63 (1940).

PART II

PRACTICAL CONSIDERATIONS
OF
THE DETERGENT PROCESS

"*In order to introduce a new idea into the mind of man, it is generally necessary to eject an old one. To move in new furniture, one has first to move out the old. All through the history of science we find that ideas have to force their way into the common mind in disguise, as though they were burglars instead of benefactors of the race.*"

— Dr. E. E. Slosson

10

Soils and Soiling

Literature references to the nature of soils and the manner in which they are attached to fabrics are relatively lacking. As one reviews the past work on detergency it is apparent that, all too often, insufficient consideration has been given to the nature of the soil that is to be deterged from a surface and even to the nature of the surface itself. Yet, a proper understanding of soils and soiling is essential to a thorough understanding of the over-all detergent process.

Nature of Soils

It is readily recognizable that soil is generally a heterogeneous mixture of many substances of different physical and chemical characteristics. Thus, Mees[1] points out that the impurities in dirty wash belong to three classes: water-soluble or polar substances; insoluble or inactive substances, mainly hydrocarbon oils; and earthy inactive substances. Rhodes and Brainard[2] suggest that the most common components of ordinary soil are probably carbon (soot and lampblack), fatty substances (from perspiration and grease), and oils.

It is evident that most past considerations of the nature of soils have been based on supposition both as to the kinds of soil and as to the probable extent to which each kind may be present on laundry work. Carrying such supposition further, the following classification of soils (exclusive of stains) is possible:

A. Water-soluble organic and inorganic material.

1. Sugars, syrup, starch, flour, and urea.

[1] Mees, R. T. A., *Z. deut. Öl-Fett-Ind.*, **42**, 235–7, 250–2 (1922).

[2] Rhodes, F. H., and Brainard, S. W., *Ind. Eng. Chem.*, **21**, 60–8 (1929).

2. Organic acids, as fruit acids.
3. Albuminous material, as blood, mucous, and egg white.
4. Inorganic materials, such as salt and lime.

B. Water-insoluble inorganic material.
1. Cement, plaster, soot, and lampblack.
2. Earthy material, such as clay and silt.

C. Water-insoluble inert organic material.
1. Hydrocarbon oils, such as lubricating oil and grease, fuel oil, road oil, asphalt, and tar.
2. Paint and varnish.
3. Inert fats, such as the greater part of animal and vegetable fats.

D. Water-insoluble reactive organic material.
1. Particularly fatty acids, such as are present to some extent in fats and in perspiration.

Certainly it is to be expected that the particular soils and the extents to which they are present in a given case will vary considerably, depending on the exposure. Thus, there will be wide variation between the soils on a mechanic's uniform and those on a dress shirt or table linen. However, in spite of the fact that literally hundreds of kinds of soil are theoretically possible, it is indeed fortunate that they may be classified into only a few groups, based on those physical and chemical properties which determine their susceptibility to "attack" by detergent solutions. These groups correspond roughly to the four listed above. Stains are purposely excluded from the present discussion because of the fact that their removal frequently requires special chemical treatment not included in normal detergent processes.

From the standpoint of detergency, the significant property possessed in common by the *soils in Group A* is their solubility in water. Water is the only detergent necessary to effect their removal from a fabric, although surface-active material in the water may be a contributing factor through such assistant action as increasing the wetting action of the water. Further, such soils present no problem of redeposition, provided adequate rinsing is employed.

The problem of deterging becomes more complex in the case of *Group B soils*, wherein the soils are neither soluble in water nor in commonly used detergent solutions. They are, as a group, solids in various states of subdivision. Their chemical activity with respect to detergent agents is limited to adsorption reactions such as those which determine whether they are hydrophilic or hydrophobic, the extent to which they may adsorb detergent on their surface, and the charge assumed by their particles in suspension. Such chemical inertness necessitates physical or mechanical action for their separation from fabric and for their dispersion in the detergent solution.

As with Group B soils, the *soils in Group C* are susceptible to neither solution in nor chemical attack by detergent solutions. They are liquid or semi-liquid substances and, from the standpoint of detergency, are similar in many respects to the solid soils of Group B. Here again successful deterging depends on physical and mechanical action. The main differentiation between soils of Group B and of Group C lies in the difference in their physical states. The deterging of Group C soils involves emulsification in addition to the actions involved for the deterging of solid materials.

The *soils of Group D* are relatively exclusive with respect to number but are none the less important from the standpoint of detergence. They are insoluble in water alone. However, in the presence of alkali from hydrolysis of soap or added as builder, they may be converted chemically to water-soluble material, the disposal of which is much the same as that of Group A soils.

Nature of Fibers

The manner of attachment of soil to fabric is closely associated with the characteristics of the individual textile fibers. Certainly the physical state of the surface and interior of a fiber and the existence of any tendency for chemical combination between fiber and soil must receive consideration in our present discussion. A brief review of the common fibers therefore is necessary. Detailed discussions of the fibers are ably presented by Matthews[3] and by Olney.[4]

The vegetable fibers, of which cotton and linen are by far the most important to launderers, are composed almost entirely of cellulose, a material which is rather inert chemically and not readily attacked by dilute acids or alkalies. The animal fibers, such as wool

[3] Matthews, J. M., "Matthews' Textile Fibers," 5th ed., New York, John Wiley & Sons, Inc., 1947.

[4] Olney, L. A., "Textile Chemistry and Dyeing," Part I, "Chemical Technology of the Fibers," Lowell, Mass., Lowell Textile Associates, 1945.

and silk, are entirely different chemically from the vegetable fibers. Proteins, in contrast to cellulose, constitute a considerable portion of the animal fibers and impart to these fibers a decided tendency to react with both acids and alkalies. This tendency is well illustrated by the adverse action of alkali on woolens. Rayons are similar in some respects to vegetable fibers in that they consist primarily of cellulose (regenerated) or of compounds derived from cellulose (cellulose acetate). Their chemical behavior is different, however, from that of

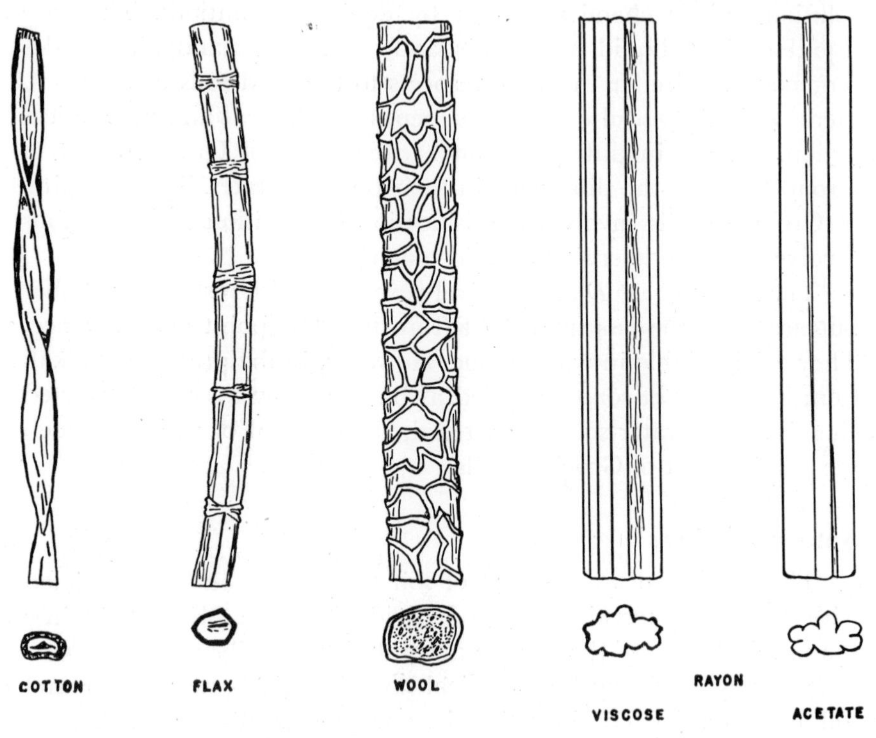

Figure 10–1. Sketches illustrating general physical structure of common textile fibers.

cotton as indicated, for example, by the difference in dyeing characteristics of either type of rayon and of, say, cotton. Nylon is quite similar chemically to the animal fibers.

The long tubular fibers of the growing cotton collapse and twist upon ripening to produce fibers which appear under the microscope as typical flattened, spirally twisted tubes. They have the appearance of twisted ribbons with thick and rounded edges, except for their centers being hollow (see Figure 10–1). In the case of mercerized cotton, the natural fibers have been changed by strong caustic alkali to straight, rod-shaped fibers whose walls have thickened to such extent as to almost eliminate the central opening or *lumen* and, likewise, the spiral.

The outer or *primary* wall of the raw cotton fiber is coated with a film of wax, pectic material, and mineral matter which is removed

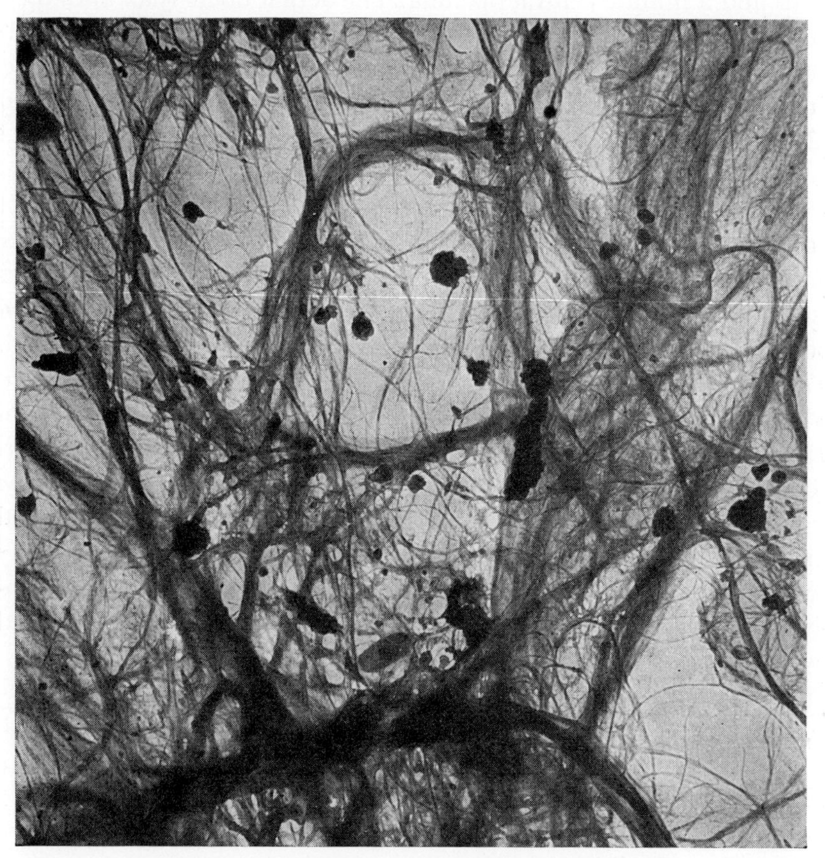

Figure 10–2. Photomicrograph of mature cotton fibers mechanically disintegrated in water, showing fibrils. (Photograph courtesy American Cyanamid Company.)

to some extent during processing into thread. The primary wall proper appears to be composed of entwining "sub-fibers" or *fibrils.* Cotton fibers are of the order of width of 15 to 20 microns, whereas the diameter of the fibrils is in the range of 0.3 to 1.0 micron. The fact that the cotton fiber is essentially a "bundle" of much smaller fibers is substantiated by observation at high magnification of mechanically disintegrated fibers, as illustrated in Figure 10–2. Our knowledge of the permeability of such "bundles" with reference to solids is limited; however, it is doubtful that solid soil of greater than submicroscopic size can penetrate the interior deeply.

A single flax fiber consists of a long transparent tube with thick walls and a minute central opening. The actual fibers (flax line) used in textiles consist of bundles of such single fibers, the number depending on the extent of hackling (combing) of the flax and the desired fineness of texture of the textile. Flax line may vary in diameter from 120 to 250 microns, whereas the single fibers may vary from 12 to 25 microns.

Wool fibers always consist of two and sometimes three distinct portions. The most evident of these under the microscope is the outer horny or scaly covering somewhat similar in appearance to fish scales (see Figure 10–1). Within the protective scaly exterior lies the *cortical* body of the fiber, from which the strength and elasticity are derived. The cortex consists of striated, spindle-shaped cells, the striations in which are due to fibrils, which in turn are composed of microfibrils. Thus, within the scaly surface, the wool fiber, like the cotton fiber, consists of a bundle of many smaller fibers. A third portion of the wool fiber, the *medulla*, lies in the center and is less extensive in good than in poor fibers. It consists of marrow-like material and probably comprises a spongy passageway for fluids of the living fiber. An extensive microscopic study of the structure of wool fibers has been reported by Hock, Ramsay and Harris.[5]

The porosity of wool within the scaly covering may be quite comparable to that of cotton, although this is not definitely known. The well-known difference between ease of washing wool and ease of washing cotton is probably associated to some extent with the scaly exterior of the wool which may reduce the permeability and, thus, the penetration of soil into the fiber.

Raw silk fiber, as "extruded" by the silkworm, consists of twinned triangular fibroins or *brins* held together by gum material. Degumming during processing frees the brins from each other. It is difficult, even under high magnification, to distinguish any cellular structure

---

[5] Hock, C. W., Ramsay, R. C., and Harris, M., *J. Research Natl. Bur. Standards*, **27**, 181–90 (1941).

in silk brins; however, it is likely that they consist of bundles of very fine fibrils which become somewhat fused together as they are excreted by the silkworm. Such fibrils are more evident in wild than in domestic silks. Interior ducts are absent. The permeability of silk fibers with reference to solids is probably very low.

Rayons, like silk, are continuous fibers. However, fibrils and a consequent "bundle" structure are absent, as well as any interior passageway. In cross section, the acetate fibers have several smoothly rounded lobes, whereas viscose rayon shows edges which are almost serrated, as illustrated by the sketches in Figure 10–1.[3] The interiors of rayon fibers probably are nearly impermeable to solid soils.

Nylon is a continuous, cylindrically shaped fiber without tubular or fibrillar construction. Its smooth exterior surface should have a significant effect in minimizing the adherence of surface soil.

Particularly worthy of special consideration are the water absorption characteristics of the various fibers, as a possible clue to penetration of water-borne soils. Natural fibers in the raw state are "water proofed" by various substances, notably gums in the case of cotton, linen, and silk, and fatty material in the case of wool. During processing of the natural fibers to threads or yarns, much of these water-repellent materials is removed.

Processed cotton fiber is particularly wettable by water. It is hydrophilic because of the many exposed hydroxyl (—OH) groups in the cellulose molecule. The wettability of linen fibers is dependent on the processing of the linen. In many cases, considerable amounts of the water-repellent gums are left on the fibers for specific reasons. Cotton fibers may swell as much as 40 per cent in volume upon immersion in water, practically all of the increase being in the cross section. The amount of water imbibed by cotton, in comparison to that imbibed by other cellulosic fibers, is shown in Table 10–1 as reported by Matthews, based on studies by Boulton and Morton.[6]

Wool fibers are strongly hygroscopic to water vapor. Even wool in the grease may absorb as much as 40 to 50 per cent of water under exposure to 100 per cent relative humidity. Some of this hygroscopicity may be attributable to the presence of hygroscopic non-protein material in the fibers. Upon immersion in water, wool fibers may swell cross-sectionally by approximately 10 per cent but return to their original size upon drying. Silk is readily wetted by water.

[6] Boulton, J., and Morton, T. H., *J. Soc. Dyers Colourists,* **56,** 145–59 (1940).

TABLE 10–1. WATER IMBIBITION BY CELLULOSIC FIBERS

Fibers	Grams Water per 100 Grams Cellulose
Viscose rayon and staple fiber	85–105
Cuprammonium rayon	90
Cotton, American	42
Cotton, American, mercerized loose	56
Cotton, American, mercerized, with tension	46
Ramie	42
Linen	46

Reproduced by permission from "TEXTILE FIBERS—5th Ed." by J. Matthews and H. R. Mauersberger, published by John Wiley & Sons, Inc., 1947.

Viscose rayon, having retained its cellulosic hydroxyl groups, shows a wettability comparable to that of natural cellulose. Acetate rayon is somewhat less wettable than viscose or natural cellulose because of substitution of acetyl groups for some of the hydroxyl groups. The cross-sectional swelling of rayons upon immersion in water has been studied by Herzog[7] and by Lawrie,[8] with the results shown in Table 10–2. The swelling of the viscose and cuprammonium rayons is particularly pronounced.

TABLE 10–2. CROSS-SECTIONAL SWELLING OF RAYONS IN WATER

Kind	Country of Origin	Percentage Increase in Area of Air-Dry Cross-Section	Reference
Cuprammonium	Germany	61.8	Herzog[7]
Viscose		65.9	
Acetate		5.7	
Cuprammonium	England	53.0	Lawrie[8]
Cuprammonium		41.0	
Viscose		35.0	
Viscose	Germany	52.0	
Acetate	England	9.0	
Acetate		11.0	
Acetate	France	14.0	

Reproduced by permission from "TEXTILE FIBERS—5th Ed." by J. Matthews and H. R. Mauersberger, published by John Wiley & Sons, Inc., 1947.

Nylon is readily wetted by water but shows low water absorption. The remarkable ease of drying nylon is evidence of this latter fact.

---

[7] Herzog, A., "Die mikroscopische Untersuchung der Seide und der Kunstseide," Berlin, J. Springer, 1929.

[8] Lawrie, L. G., *J. Soc. Dyers Colourists*, **44**, 73–8 (1928).

Mechanics of Soiling

Past studies of the mechanics of soiling have been limited almost entirely to soiling as employed in connection with standard test pieces for detergency tests. Two criteria have, in many cases, determined the soiling techniques used for such studies: (1) the reproducibility and uniformity of the soiling; (2) the ease of quantitatively measuring the soil before and after washing. Experience gained from such studies, while affording much help in evaluating detergents, is of little or no value when considering the mechanics of soiling under "natural" conditions.

In the practical absence of laboratory studies of the mechanics of soiling, it is necessary to resort to postulations of the parts played by various factors in the soiling process. We do not know all the factors involved but certainly the following play significant parts, both in the extent to which a soil penetrates a fabric and in the tenacity with which it is held:

(a) *Characteristics of the soil.* The manner in which soil may enter into the interstices of a fabric should be dependent to a considerable extent on whether the soil is wet or dry and, if wet, whether the liquid is water or an oil. Dry soil, such as earthy material, will not readily penetrate of its own accord. In the case of wet soil, the penetration of the liquid phase by capillary action will carry solid material into the interstitial capillaries of the fabric. Further, a hydrophilic solid material should penetrate more readily than a hydrophobic solid material when wetted by water, because of the fact that, in the former case, water is a mutual wetting agent for both the fabric and the soil. On the other hand, there are solid soils which are more readily wetted and dispersed by oils than by water, in which case oil as the "vehicle" will effect the greater penetration of solid material into the fabric. These various conditions of the soil and the part they may play in soiling are illustrated in the photographs of Figures 10–3 to 10–6.

It may be seen from Figure 10–3 that penetration of fabric by *dry* solid soil is dependent on attendant mechanical action. On the other hand, as shown in Figures 10–4, 10–5, and 10–6, penetration by *wet* solid soil is dependent to a considerable extent on the wetting characteristics of the solid. Figure 10–4 shows the greater penetration of clay when suspended in water than when suspended in kerosene, presumably because the clay is more readily wetted and dispersed by water. Figure 10–5 shows that *activated* charcoal is about equally wetted by water and kerosene, with the result that there is no pronounced difference in penetration. In the case shown by Figure 10–6,

Side of cloth on which soil was applied

Reverse side of cloth

Not Rubbed Rubbed

Figure 10–3. Photographs showing penetration, due to rubbing, of dry clay through cotton fabric.

the situation is completely reversed. The soot, being somewhat oily, is hardly wetted by the water but readily wetted by kerosene. Penetration of the water-suspended soot is only slight, whereas penetration of the kerosene-suspended soot is very pronounced.

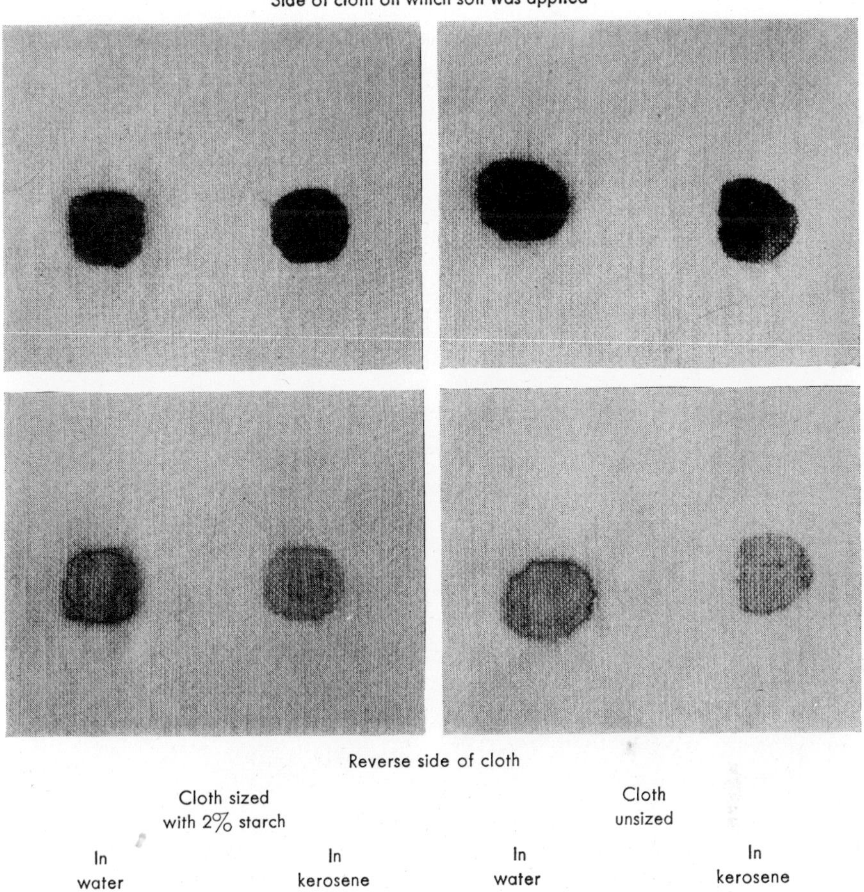

Figure 10–4. Photographs showing penetration through cotton fabric by clay suspended in water and in kerosene.

(b) *Conditions of soiling*. Any mechanical rubbing action during soiling will tend to "grind" solid soil into the fabric, even in the absence of a liquid vehicle. An example of this condition faced every day by laundrymen is the rubbing of soot or dust into shirt sleeves and collars. A further factor to be considered under the conditions of soiling is whether soiling is by contact or by immersion. For the present case, immersional soiling will be defined as soiling under any condition where the fabric is completely wet through by the liquid phase of the soil. The parts played by these various conditions also are illustrated in Figures 10–3 to 10–6.

Side of cloth on which soil was applied

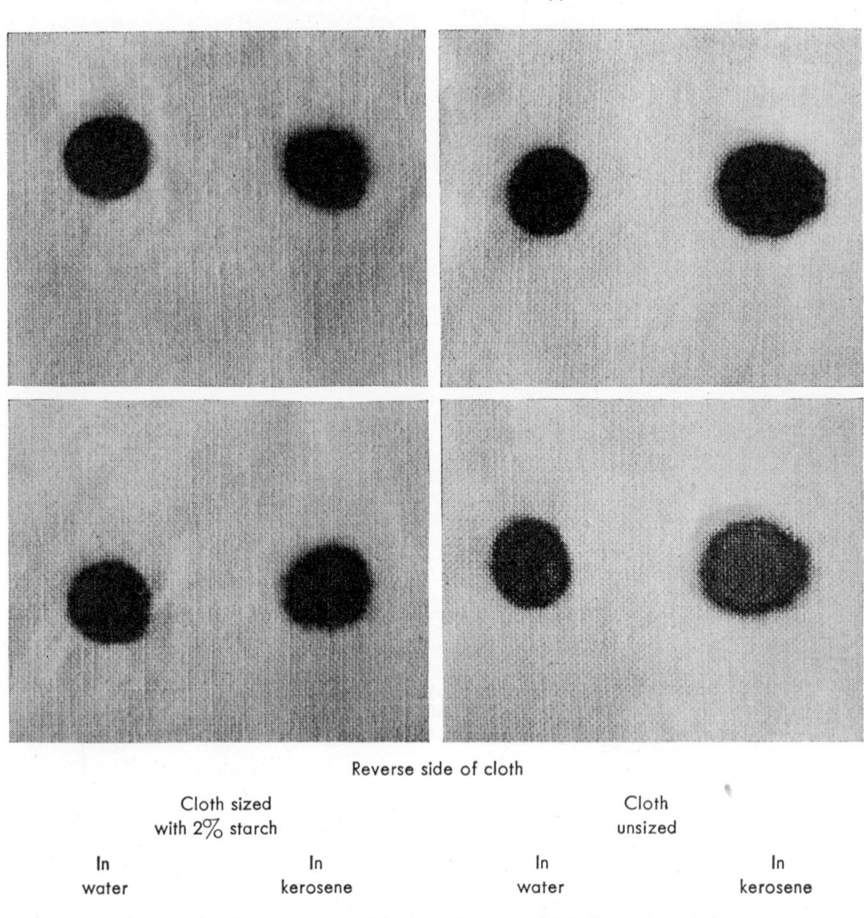

Reverse side of cloth

Cloth sized with 2% starch		Cloth unsized	
In water	In kerosene	In water	In kerosene

Figure 10–5. Photographs showing penetration through cotton fabric by activated charcoal suspended in water and in kerosene.

(c) *Characteristics of the fabric.* Other factors being equal, some fabrics are more readily penetrated by soil than are others. The kind of fibers in a given fabric, and its particular wettability, surface "roughness," and interior permeability will play a part. The manner in which the fibers are spun and woven into the fabric should be of much significance. Thus, fabrics loosely woven from coarse threads should be more readily penetrated by solid soil than finely woven fabric. Fabrics, such as toweling, which are designed for ready pene-

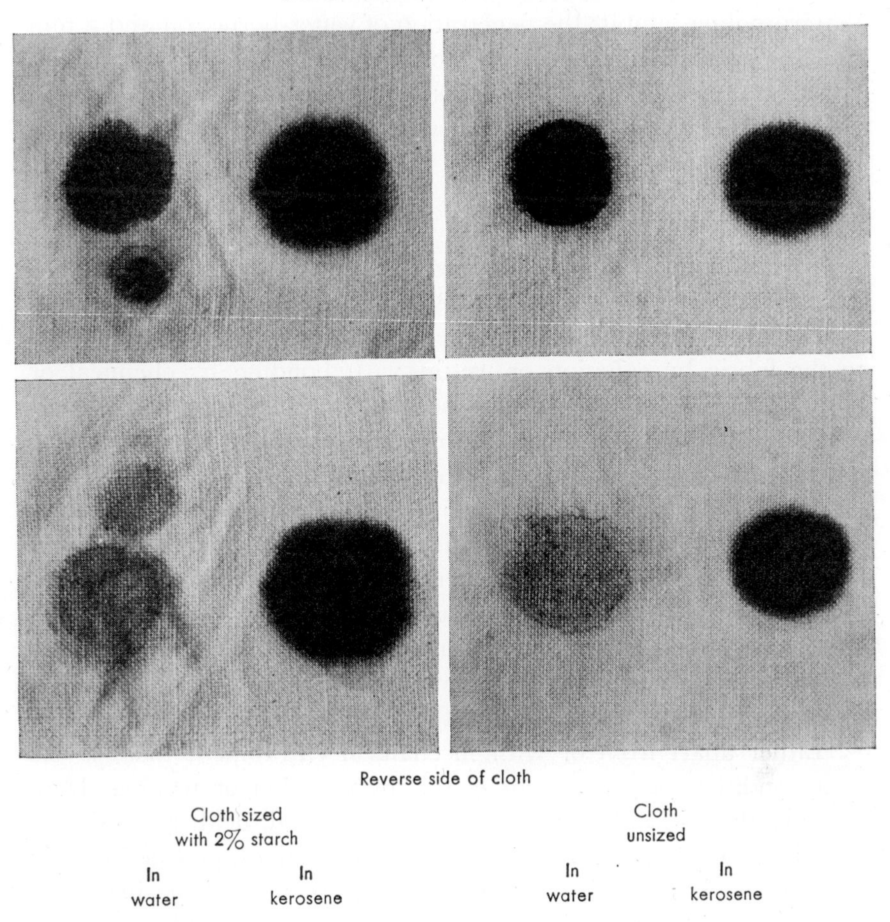

Figure 10–6. Photographs showing penetration through cotton fabric by soot suspended in water and in kerosene.

tration by and high absorption of liquids may be readily and extensively penetrated by soil.

(d) *Condition of the fabric.* The "wetness" of a fabric prior to soiling should determine part of the penetrability of the fabric by soil. Thu the swelling of all common fibers when wet by water increases their permeability by water-borne soil while, on the other hand, the presence of water in such cases may inhibit the penetration of oily material. Further, more specific cases of "soil proofing" are those obtained with fabrics which have been purposely waterproofed or with fabrics which have been sized by such a material as starch. Waterproofing inhibits the penetration of water-borne soil and a film of sizing agent on and within a fabric may well inhibit the penetration of any soil. The latter case is well illustrated by the common experience of the assistant action of starch in the washing of such garments as shirts.

Manner of Attachment of Soil to Fabric

Several conditions undoubtedly contribute to the attachment of soil to fabric, all of which probably fall under one or another of the following: (1) mechanical entrapment on or in the fabric; (2) bonding to the fabric by cohesion or wetting; (3) bonding by chemical or adsorptional combination with the fabric. On a thorough understanding of these conditions depends the final answer as to *how* a detergent solution accomplishes the separation of the soil from the fabric.

Soil may be present as a liquid, a solid, or a mixture of liquid and solid, an example of the latter being soot in oily material. Mechanical entrapment can be limited to solid soil alone where there is no true bonding but, rather, entanglement of the solid particles either within the mesh of the fabric, between the individual fibers, or on the rough protuberances of the fiber surfaces.

Either apart from or with mechanical entrapment there is the next condition of attachment of soil by cohesion or wetting. Here we have the case of the soil, if it be only a liquid, held by wetting of the fiber surfaces or, if it be liquid and solid, the liquid acting as a bonding agent between solid soil and fabric by mutual wetting. Such a condition of soiling is analogous to the soiling of a china plate by grease. However, in the case of fabrics, the condition becomes more complicated because of the multitude of capillaries with attendant large surface areas and restricted passageways. The strength of the bond in these cases becomes a function of the energy required to destroy the interface between liquid and fiber surface. Destruction of the bond requires either an increase in the free energy of the liquid soil-fabric interface, the substitution of a new interface of lower free energy (such as a soap solution-fiber interface), or the imparting of mechanical energy to the soil-fabric interface as by rubbing.

The significance of the third condition listed above, chemical or adsorptional combination of the soil with the fabric, depends upon the definition we wish to assign to the term **chemical combination.** With few exceptions, such as with strong acids or alkalies, we do not think of most soils as being capable of combining chemically in the ordinary sense with cellulosic or proteinaceous material of fabrics. However, in a broader sense, we may consider the action of adsorption as a chemical or pseudo-chemical combination. Spring,[9-12] one of the early investigators of detergency, considered fine carbon particles in water to be capable of forming what he termed a combination of adsorption more or less stable with solid substances such as cellulose. Adsorption in soiling is particularly significant, not because it necessarily plays a greater role than the other two conditions already discussed, but because it is so complex and little-understood.

In discussing the process of adsorption as it might pertain to soiling, the inclusion of all possible soil combinations is not practical. The present discussion will be limited to two general types of soil combinations, the characteristics of which are as follows:

(a) A solid-liquid combination wherein the liquid is preponderantly water and which contains some "polar-nonpolar" organic material as well as inert solid or liquid material, the "polar-nonpolar" material possibly being fatty acid or any other oily material that contains a hydrophilic group;

(b) A solid-liquid combination wherein the liquid is preponderantly oily material, but which is otherwise similar to the above combination.

It is considered that much of the soil that is encountered in laundry work will fall into one or the other of these two general types.

As has been pointed out by Glasstone[13] in the case of adsorption by solids from solution, "the function of the surface may be con-

---

[9] Spring, W., *Rec. trav. chim.*, **28,** 120–35 (1909).
[10] Spring, W., *Z. Chem. Ind. Kolloide,* **4,** 161–8 (1909).
[11] Spring, W., *Bull. acad. roy. Belg.*, **1909,** 187–206.
[12] Spring, W., *Rec. trav. chim.*, **28,** 424–43 (1909).
[13] Glasstone, S., "Textbook of Physical Chemistry," 2nd ed., p. 1215, New York, D. Van Nostrand Company, Inc., 1946.

sidered from two aspects: it may be regarded as acting in the same manner as in the adsorption of gases, i.e., by virtue of molecular attraction or of chemical forces, or it may be looked upon as a means for providing an interface of large area at which a solute capable of lowering the interfacial tension may accumulate. These two properties are undoubtedly related to each other, but the latter viewpoint is probably more useful; . . ."

Turning now to the first type of soil combination listed above, the soil may be held to a fabric by all the conditions of mechanical entrapment, cohesion and wetting, and adsorption. The adsorption and even the wetting, in turn, may be influenced by surface-active material present in the soil. The surface-active material may be fatty acids or any other material which contains in its molecules contrastingly hydrophilic and hydrophobic groups. The presence of such material in the soil will result in its positive adsorption at the soil-fabric interface, with attendant lower interfacial free energy and greater tenacity of bond. The conditions obtaining at the interface in this case are schematically illustrated in *A* of Figure 10–7.

The case with a soil combination of the second type, wherein the liquid phase is predominantly oil, is no different basically from that just described. The only variation is that, in an oil as the solvent or continuous liquid phase, a surface-active material is oppositely oriented from what it is in water. Positive adsorption of polar-non-polar solutes occurs in oil as well as in water. The case is now illustrated as in *B* of Figure 10–7, and here again positive adsorption of a surface-active soil constituent may play a significant part in the "bonding" of the soil to fabric. In either case, the influence of the surface-active soil renders subsequent soil removal more difficult. It permits the soil to "wet" the fabric more thoroughly, with the result that a wetting combination of still greater effectiveness is required in order to deterge the soil from the fabric.

The case for true chemical combination between soil and fabric is probably limited to proteinaceous fibers and to any acidic or basic constituents in soil. The possibility of chemical reaction between a fatty acid and a basic group in, say, a wool fiber is conceivable, provided that dissociation of the two reactants is sufficiently pronounced to open the way for the reaction. Undoubtedly, water must be present if such reaction is to take place. Combinations of this kind, between both acidic and basic constituents of soil and fabric, lack study.

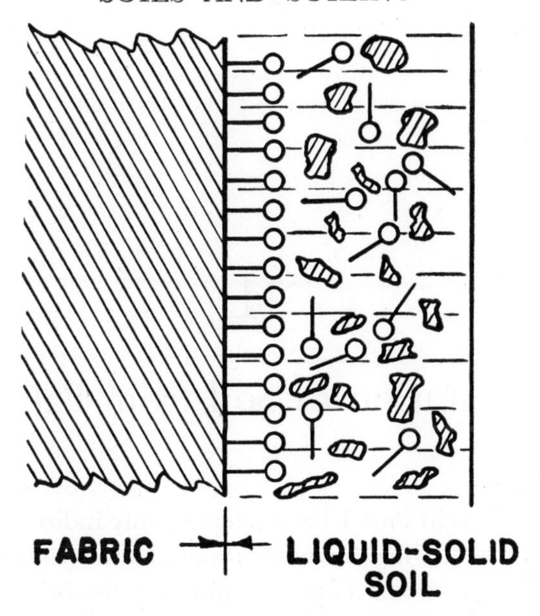

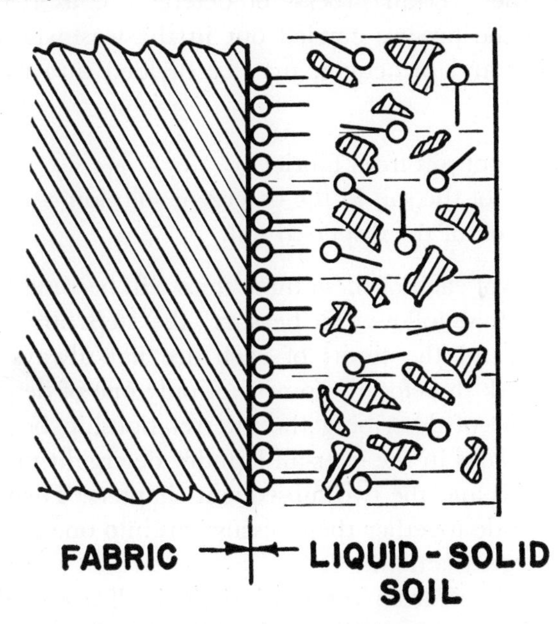

Figure 10–7. Illustration of the role of surface-active soil (such as fatty acids) in the bonding of soil to fabric.

11

Separation of Soil from Fabric

The discussions in Part I have referred only indirectly to practical detergent processes. They have been intended to constitute the theoretical foundation upon which to build our discussions of the more practical aspects of detergency in laundering.

Basically, the over-all process of detergency, as represented in laundering by the process carried out in the washwheel up to the point of rinsing and bleaching, may be divided into three steps or "sub-processes":

a. Separation of soil from fabric;
b. Dispersion of the soil in the detergent solution;
c. Stabilization of the dispersed soil.

The significance of each of these three steps will become evident from the discussions that follow. To be sure, the steps are considerably interdependent and the limits of each are not always clearly delineated. Our task therefore becomes not only one of describing each step but also one of describing the interrelationship between steps. The individual steps in the over-all laundry detergent process will be considered in this and the two subsequent chapters. Then an attempt will be made to tie together these discussions into one comprehensive picture of detergent action.

Separation of soil from the fabric is in itself a complex subject. Those soils which are readily and completely water-soluble constitute no problem in detergency. Their separation by the ordinary dissolving action of water, although of great significance in the laundry washing process, warrants no particular detailed consideration because of its technical simplicity. Water-insoluble inorganic material may be considered chemically inert in so far as detergent

Figure 11–1. View of operating laundry department of the American Institute of Laundering.

solutions are concerned and require for their separation such physical actions as can be imparted by the detergent solution plus, generally, applied mechanical agitation. Water-insoluble inert organic materials are much the same in this respect, except that they are most frequently liquid or semi-liquid. Their separation from fabric involves physical actions, some of which are of a different nature than those required for the predominately solid inert inorganic soils. Finally, the separation of water-insoluble reactive organic materials involves, in addition to physical action, chemical conversion of the soil to other material by some constituent in the detergent solution.

The known actions of detergent solutions which determine their ability to effect or assist in the separation of soil from fabric are several; namely, solution of water-soluble material, neutralization of acidic material, wetting and spreading on both fabric and soil surfaces, lowering of the free surface energy of soil surfaces, enhancement of pedesis of solid and liquid particles, foaming, and solubilization. The theoretical considerations of these actions have been covered in preceding chapters.

In one of the earliest attempts to account for the separation of soil from fabric by soap solution, Spring[1-4] postulated a pseudo-combination of soil with fabric which is replaced by a somewhat more stable combination of soil with soap. In the sense of adsorption combinations, Spring's postulation might be acceptable; however, it lacks much of being a complete explanation of the facts.

The neutralizing action of either the hydrolysis alkali of a soap solution or of alkali added as a builder was recognized early by such investigators as Jackson[5] and Shorter[6] as being a significant factor in soil separation. Thus, the neutralization of fatty-acid soil to form a readily soluble soap not only aids in the removal of that fatty acid soil but produces extra soap which is so situated as to be particularly effective in furthering the removal of other soil.

De Keghel[7] considered that, in the process of laundering, the soiling agents can be eliminated according to their solubilities. In this way we would expect first the removal of the readily water-soluble materials, such as sugars and water-soluble salts, and then the removal of organic acid material by reason of its conversion to

---

[1] Spring, W., *Rec. trav. chim.*, **28,** 120–35 (1909).
[2] Spring, W., *Z. Chem. Ind. Kolloide*, **4,** 161–8 (1909).
[3] Spring, W., *Bull. acad. roy. Belg.*, **1909,** 187–206.
[4] Spring, W., *Rec. trav. chim.*, **28,** 424–43 (1909).
[5] Jackson, H., *J. Roy. Soc. Arts*, **55,** 1101–14, 1122–32 (1907).
[6] Shorter, S. A., *J. Soc. Dyers Colourists*, **32,** 99–108 (1916).
[7] De Keghel, M., *Rev. chim. ind.*, **30,** 171–8 (1921).

water-soluble alkali salts. He further considered that when the remaining soiling agents no longer react with the detergent solution, the detergent acts on the support rather than on the soiling agent itself. The deterging solution is "soaked up" by the supporting medium (the fabric), adheres to it, undermines the zone of contact of the soiling agent, and finally loosens it from the support so that it is readily carried away from the fabric mechanically.

Preferential adsorption of the detergent on the fabric, to the exclusion of such adsorption on the soil, as postulated by de Keghel, is not readily conceivable. There is every reason to expect that the detergent is readily adsorbed on the soil surface; otherwise, we could not realize the pronounced lowering of the free surface energy of, say, oily soil, which aids so much in emulsification. On the basis that soil is held to the fabric by adsorption, Guernsey[8] considers that the detergent must be adsorbed more strongly by the soil than the soil is by the fabric, thus, in effect, displacing the fabric. It is probable that the adsorption of detergent both on the soil and on the fabric plays a very significant role in the separation of the soil from the fabric. Further, Mikumo[9] points out that, in the case of a mixture of a homologous series, the higher soap is always selectively adsorbed at any interface.

The mechanics of the penetration of fabrics by a detergent solution have been well described by Powney.[10] He points out that penetration of the various capillaries of the fabric is attended by displacement of air from the capillary spaces. In the case where detergent solution is converging on the center of a "bundle" of fabric from all directions, there is a specific tendency, in the absence of mechanical agitation, for entrapment of air and imperfect penetration of the solution, particularly into the finer capillaries. The air pressure in such a case may become sufficient not only to overcome the gravitational forces which account for *immersional* wetting but also to inhibit the *spreading* wetting occasioned by the surface activity of the solution. Powney also points out a condition whereby the detergent itself may inhibit penetration, as occasioned by the "soaking-up" of water by the fibers to such an extent that a resulting residue of detergent gel

---

[8] Guernsey, F. H., *Am. Dyestuff Reptr.*, **12**, 766–8 (1923).

[9] Mikumo, J., *J. Soc. Chem. Ind.*, **52**, 65–68T (1933).

[10] Powney, J., "Wetting and Detergency Symposium," pp. 185–96, Brit. Sect. Intern. Soc. Leather Trades' Chem., 1937.

in very fine capillaries may block further penetration into these capillaries.

The pronounced wetting and spreading powers of detergent solutions, as compared to those of water alone, result in more rapid and extensive penetration by the solution into the interstices of both the fabric and the soil. The stronger adsorption of detergent than of soil on the fiber surfaces loosens the bond between soil and fiber. At the same time, adsorption of the detergent on the soil surface results in a marked reduction in surface energy of the soil. If the soil is a solid material, the loss of its bond with the fabric permits it to be swept away in the detergent solution by pedesis or by mechanical agitation. If, in addition, the soil is an aggregate of solid particles, the adsorption of detergent not only on the surface of the aggregate but in its interstices may accomplish the peptization of the aggregate into individual particles which may more readily find their way through the voids of the fabric and into the solution. If the soil is a liquid, oily material, the reduction of its interfacial tension greatly facilitates its emulsification by mechanical action. Further, if an oily soil contains free fatty acid, neutralization of the fatty acid to a soap may even contribute to the more or less spontaneous emulsification of the inert oily portion as previously discussed in Chapter 5. Finally, the film formed in a detergent solution at the interface between the solution and an air bubble (as in foaming) furnishes a seat for adsorption of solid or liquid soil particles so as to aid in carrying away the particles from the fabric.

It may be seen from the foregoing discussion that a very significant role of a detergent in separating soil from fabric is that of "readying" the soil so that it may be more easily removed by some outside action such as mechanical agitation. To summarize, it loosens the bond between soil and fabric, it reduces the free surface energy of the soil, it peptizes solid aggregates into individual particles, and to some extent it may effect spontaneous emulsification and assist spontaneous separation by pedesis.

We have already mentioned solution of certain soil materials by the detergent solution, which is a complete action of removal accomplished by the solution itself. One other independent action of the detergent solution, namely, solubilization, also accomplishes complete removal. As has been mentioned previously, the significance of the extent of solubilization of insoluble material in detergent solutions of the concentrations used in the washwheel is questionable. Most of the studies of solubilization in the past have been made with detergents at concentrations in the range of 5 per cent or more, whereas washwheel concentrations are in the range of a few tenths of one per cent. To be sure, the detergent micelles in 0.2 per cent solution can be expected to solubilize some material but even under equilibrium conditions the extent of this action is measured in terms of only a very few milligrams of solubilized material per 100 cc of solution. In this connection Preston[11] has pointed out that, whereas washing power of a soap solution appears to reach a maximum at about the critical concentration for micelle formation, solubilization apparently only begins at this concentration. On the other hand, the special application of "spotting" in laundering, whereby soap is rubbed on spots that are particularly heavily soiled, presents a different situation. In such applications the concentration of the soap solution on the spot may be very high and the extent of solubilization may be pronounced. The result is "incasement" of the soil within the detergent micelles to effect separation from the fabric.

Attempts to find a readily measured true index of detergent power have been very numerous but have met with only partial success. Perhaps the most persistently pursued study has been that of attempting to correlate surface or interfacial tension with detergent power. Actually, surface tension as usually considered, that is, as the tension at an air-liquid interface, is of little direct consequence in detergency except in connection with foaming and with the replacement of air from the fabric being washed. On the other hand, interfacial tension, whether it pertains to a liquid-liquid or liquid-solid interface, is closely associated with detergent power. Agents which markedly lower the interfacial tension of water against another phase increase the wettability of the other phase by the water and decrease its free surface energy. These are actions which we have just discussed as being of such importance in the process of separating soil from fabric. However, the inadequacy of interfacial tension alone as an index of detergent power is easily illustrated by the fact that there are many excellent wetting agents which are very poor detergents. Adam[12] considers that interfacial tension is probably not of fundamental importance to detergent action in the sense that it alone

[11] Preston, W. C., *J. Phys. & Colloid Chem.*, **52**, 84–97 (1948).
[12] Adam, N. K., *J. Soc. Dyers Colourists*, **53**, 121–9 (1937).

determines detergent power, nor is it necessarily the predominating factor in determining this efficiency. Interfacial tension is important, but there are other factors, less readily measured, which are at least equally important.

In a study of the manner in which oil is removed from wool fibers, Adam[12] has observed microscopically the conversion of the oil film into globules by the action of the detergent solution, after which the globules are readily separated from the fiber by mechanical agitation. From this he concludes that the contact angle formed by the detergent solution-oil interface with the fiber surface is closely related to de-

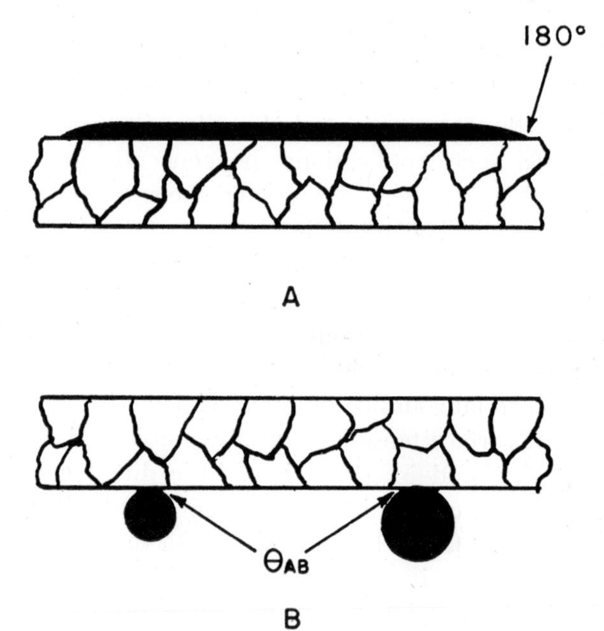

Figure 11–2. Alteration of angle of contact between a grease-water surface and a solid surface during the detergent process.[12]

tergent efficiency. This contact angle, which is illustrated by Figure 11–2, taken from his paper, is treated mathematically as follows, and incorporates discussions presented in previous chapters:

(a) For water to spread over a fiber or oil surface as a continuous film, the condition for perfect wetting is that there be no angle, measured in the water, between the water surface and the other surface. In other words, as pointed out originally by Young,[13] the

[13] Young, T., *Phil. Trans.*, **65**, (1805).

water must attract the other surface more than it attracts itself. Mathematically, the contact angle under conditions of simple wetting is expressed in energy terms as follows:

$$W_{sw} = \gamma_w(1 + \cos\theta) \tag{1}$$

or

$$\cos\theta = \frac{W_{sw}}{\gamma_w} - 1 \tag{2}$$

where W_{sw} is the work of adhesion between the water and the other surface, γ_w is the surface tension of the water, and θ is the contact angle. Since $2\gamma_w$ is a measure of the cohesion of the water itself or of the work necessary to "pull" water from water it is necessary that W_{sw} be at least equal to $2\gamma_w$ in order to realize perfect wetting. Wetting may be promoted either by increasing the adhesion of the water for the other surface, or by decreasing the surface tension of the water, or by a combination of the two. The role of soap or other detergents in promoting the wetting of an already hydrophilic surface (such as fiber surface) by decreasing the surface tension of the water and in promoting the wetting of a hydrophobic surface (such as an oil surface) by converting that surface to a hydrophilic one has already been discussed.

(b) In order to obtain the best penetration of water into a porous solid, such as into the interstices between the individual fibers of a fabric or between the individual particles of an aggregate of solid soil, it is not sufficient that W_{sw} be just slightly greater than $2\gamma_w$. It is desirable that the adhesion of the water for the solid surface be as much greater than the adhesion of the liquid for itself as possible, or, in other words, that the terms of

$$\gamma_w \cos\theta = W_{sw} - \gamma_w \tag{3}$$

be as large as possible.

(c) Probably the only factor in emulsification of an oily liquid or dispersion of a hydrophobic solid aggregate that involves energy considerations alone is that the interfacial tension between the oil or the solid and the water must be reduced as much as possible. As shown by Dupré, the interfacial tension between the water and oil is related to the surface tensions of the two liquids separately against air by the equation

$$\gamma_{wo} = \gamma_w + \gamma_o - W_{wo} = \gamma_o - (W_{wo} - \gamma_w) \tag{4}$$

and a similar equation holds for the interfacial tension between the solid and the water:

$$\gamma_{sw} = \gamma_s - (W_{sw} - \gamma_w) \tag{5}$$

Adam points out that, in equations (3), (4), and (5), the expression which must be as large as possible in order to obtain maximum penetration or ease of dispersion is the same in each case—the *adhesion tension*, $W - \gamma_w$, of the water for the oil or the solid must be as large as possible.

Based on this mathematical treatment, Adam considers that detergent action may be only a more complicated case of the same kind of phenomenon as that discussed above. Thus, from his microscopic studies of oiled wool fibers in detergent solutions, he concluded that the deterging of the oil from the fiber consisted of an alteration of the angle of contact between the oil-water surface and the solid surface being cleaned, from 180° to 0° measured in the water. Referring again to Figure 11–2, *A* indicates the oil spread in a thin layer on the fiber surface and the contact angle is 180° in the water (0° in the oil). An efficient detergent produces the condition shown in *B*, where the angle is 0° in water (180° in the oil).

Adam expresses the contact angle, θ_{ow}, made by the oil-water interface with the solid surface in energy terms, as follows:

$$\cos \theta_{ow} = \frac{W_{ws} - \gamma_w - (W_{os} - \gamma_o)}{\gamma_{ow}} \tag{6}$$

For the most efficient detergent action, $\cos \theta_{ow}$ should be as large as possible. As may be seen from an examination of the right-hand side of equation (6), an efficient detergent fills the dual role of increasing the numerator and decreasing the denominator.

Adam summarizes his theory of detergent action or, more specifically, of the separation of oil from fiber as follows:

". . . assuming that the oil and its relation to the solid (the fiber) are not affected by the detergent, which will probably be the case *at first*, unless the detergent is soluble in the oil, it is desirable to diminish both the surface tension of the water, and its interfacial tension against the oil; but it is also desirable to increase the adhesion of the water for the solid surface, in order to obtain the best results. Thus, neither surface nor interfacial tension, nor the adhesion of the aqueous detergent solution for the fiber, is fundamental in the sense that it alone is the predominating factor; but a certain combination of all three may possibly be the fundamental quantity."

From his experimental work, wherein he measured values of the contact angle of equation (6), he concluded that the detergent must so alter the relationship between the water and the fiber that the water tends to advance along the fiber with a very small angle in front to figuratively wedge the oil from the fiber. His observed values for the minimum concentrations of various detergents to give zero advancing contact angles, in the case of lanolin on wool, are shown in Table 11–1, and agree fairly well with the concentrations found to

TABLE 11–1. DETERGENT CONCENTRATIONS FOR ZERO ADVANCING CONTACT ANGLES FOR LANOLIN ON WOOL [12]

Detergent	Minimum Concentration for Zero Advancing Contact Angle, θ_{ow}, %
Soap	0.3
Cetane sodium sulfonate	0.04
Cetyl sodium sulfate	0.02
Various detergents with solubilized amide end-groups	0.01 to 0.02

give good detergent action in actual cleaning of wool. Although he did not report specific values, he found that with cotton much smaller differences between the concentrations of different detergents were required for producing zero advancing contact angles. These concentrations again checked well with those found to be effective in actual washing.

Numerous techniques have been developed for the direct determination of detergent power, none of which is without definite limitations. These techniques involve artificial soiling of fabric, washing of the solid fabric in the particular detergent solution under study, and finally estimating the extent of soil removal. The soils used are generally designed to be comparable to the actual soils encountered in laundering. However, the extent to which the composition of natural soil and the manner of soiling is duplicated in these tests is subject to some question. For example, it does not seem reasonable to select carbon black as the representative solid soil, as is so often done in these tests. Carbon black differs considerably from other common solid soils, particularly in the physical properties that are of such importance in detergency. The same can be said for those tests using manganese dioxide or burnt umber or some other material as the standard solid soil. It appears from past work that the choice of standard solid soil has been governed more from the standpoint of ease of measurement of residual soil in the washed fabric than from the standpoint of being representative of natural solid soils. Thus, no one solid soil can be representative of all of even the most common solid soils encountered in regular laundry work, and when combinations of solid soils are used for test purposes the problem of measuring residual soil after washing is greatly complicated. Another significant factor is the manner of applying standard soil, which seems to have received all too little attention in past work. Thus, a very common practice has been to dissolve the oily portion of a standard soil and disperse the solid portion in a volatile organic solvent, immerse the cloth in this solution-dispersion, and then remove the volatile vehicle by drying. Such practice cannot be considered representative of more than isolated cases of actual soiling.

The foregoing discussion is not intended to imply that previous attempts to measure detergent power directly are without value. In fact, the results of these studies have been of great value, particularly when interpreted with an understanding that they are not all-inclusive. These techniques have permitted quite accurate relative evaluations of detergents *under a given set of conditions*. However, only when these relative evaluations are made under a number of conditions can they be considered to be accurate representations of actual laundering.

12

Dispersion of Soil in the Detergent Solution

Dispersion, limited to insoluble soil, may be considered as that part of the detergent process starting when the soil-fabric bond has been destroyed and ending with the soil dispersed in one form or another in the detergent solution away from the fabric. As previously pointed out, the delineation between separation and dispersion is not distinct. Thus, in the process of destroying the soil-fabric bond, modifications of hydrophobic soils, particularly, are already under way to aid greatly in subsequent dispersion.

In considering the dispersion process it is important that we differentiate between hydrophilic and hydrophobic soils. This is particularly important from the standpoint of the energy required to effect dispersion. Any dispersion of a large particle or droplet into many small particles or droplets requires that energy be supplied to the system. Thus, starting with one particle or droplet having a certain total free surface and corresponding total free surface energy, dispersion into many small particles or droplets results in a marked increase in both the total free surface and the total free surface energy. If the original particle or droplet is hydrophilic, the free surface energies involved in the dispersion of hydrophilic particles or droplets are much smaller than those for hydrophobic particles or droplets.

The energy input required to effect the dispersion of a 1 cc spherical oil droplet to droplets of a uniform diameter of 10^{-5} cm., in the absence of another liquid phase, has already been illustrated by the calculation on page 90. The increase in total free surface energy in this particular case was from 121 ergs to 15,000,000 ergs and is an approximate indication of the external work required to effect the dispersion. It is true that an oil droplet of 1 cc volume is much larger than any encountered in laundering, but it serves to illustrate the present point. Next, suppose that the same oil droplet is to be dispersed to the same extent in water. An illustration of the change in free surface energy and, thus, of energy input that might attend such dispersion is as follows:

Starting with one spherical oil droplet, 1 cc in volume, in water, the interfacial tension between the two liquids is approximately 47 dynes per cm., when the surface tension of the oil is 25 dynes per cm. and that of the water is 72 dynes per cm. The total free interfacial energy in the system will be 47×1.24^2 (diameter) $\times 3.1416 = 227$ ergs. Dispersion of the oil droplet to droplets of a uniform diameter of 10^{-5} cm., which results in an increase of the total surface or interfacial area to 600,000 sq. cm., increases the total free interfacial energy from 227 ergs to 28,000,000 ergs.

In other words, there is a theoretically greater energy demand in this case than in that of dispersion of the oil in air.

Next, suppose that the oil droplet becomes "coated" with a surface film of detergent molecules or ions with their polar, "water-loving" ends oriented outward. In laundering, this coating of the original undispersed droplet takes place during the separation of the oil from the fiber, as previously discussed, and, in effect, results in the conversion of the oil to a hydrophilic droplet with considerably reduced total free surface energy. Assume that the interfacial tension is reduced to 1 dyne per cm. by the presence of the detergent, not only on the original droplet but also on the small droplets as they are formed. By the same mathematics as that used above, the energy increase involved during dispersion is now from 4.8 ergs to 600,000 ergs. Thus, by the presence of the detergent, the theoretical energy input is decreased by a factor of $\frac{1}{47}$.

We now may refer again to our previous statement differentiating between hydrophilic and hydrophobic materials in the dispersion process. Those soil materials which are naturally hydrophilic or which have been converted to a hydrophilic character by adsorbed detergent are the most readily dispersed by, say, mechanical agitation during the washing process. The advantage of the use of soap or some other detergent in this respect is obvious.

When we think of dispersion we usually think of mechanical action in one form or another. In the conventional laundry washwheel, the mechanical agitation imparted by the wheel, the "flushing" of the detergent solution through the fabric, and the rubbing and bumping imparted by the work itself are certainly essential to the attainment of the required degree of dispersion of the soil. It is not the purpose of the present dissertation to discuss the mechanics of washwheels. Rather, we are more concerned with those dispersing actions taking place in laundering which are quite independent of the mechanical action of the wheel.

Peptization, in the broad sense used in Chapter 5, can conceivably account for dispersion of soil in the absence of mechanical agitation. Much of the benefit of pre-soaking of soiled work, when used, may be attributable to this action. Certainly albuminous or other gelatinous soils are susceptible to peptization by water alone, that is, water alone may disperse them from a "gel" to a "sol" state in time, without benefit of agitation. Further, aggregates of solid particles, if the individual particles are sufficiently small, may be peptized and dispersed effectively if the peptizing agent imparts to the particles a sufficient "solution pressure." Fuchs'[1] previously discussed requirement for effective deflocculation of solid particles in water (page 97),

$$A_{pp} < A_{pw} = A_{ww}$$

appears entirely applicable to the present case of peptization. Thus, again, if the soil is highly hydrophilic or can be rendered so by adsorption of a detergent, peptization may enter into the picture of dispersion in laundering to an appreciable extent.

Several observations have been made of pedesis (Brownian movement) as a possible significant action in dispersion. Most notable of these are the observations by Jackson[2] of solid soil particles on linen fibers in distilled water and in different solutions. He observed that in the water a few particles separated from the fibers and dispersed into the water by movements of the usual oscillatory character, and that when soap solution was substituted for the water the extent of this action was considerably increased. On the other hand, substitution of a salt or of a too strongly built soap solution tended to

[1] Fuchs, *Exner's Repert. Physik.*, **25**, 735 (1889).
[2] Jackson, H., *J. Roy. Soc. Arts*, **55**, 1101–14, 1122–32 (1907).

arrest pedesis. Jackson points out, however, that soap does not itself enhance pedesis but, rather, overcomes factors, such as agglomeration, which inhibit the normal functioning of pedesis. Energy considerations support this point. Pedesis is due to "bombardment" of the particles by molecules of the suspending medium, and soaps or other detergents certainly do not accelerate these molecules. Jackson considered this "aiding" of pedesis by soap to have considerable potential value in washing, particularly of materials whereon vigorous mechanical agitation must be avoided. As with peptization, its place of greatest value is probably in "pre-soaking." In the washwheel, dispersion by pedesis can be no more than a very small fraction of the mechanical dispersion.

Along with peptization and pedesis, brief mention must be made of spontaneous emulsification. As was stated in Chapter 5, the action of "shattering" dispersion of oily material has been observed on occasion when fatty acid in the oily material was neutralized by the alkali of the detergent solution. The spontaneity of this action was checked by careful control against outside agitation. However, much remains to be done to determine both the optimum conditions and the significance of this action in laundering (see also p. 241).

The action of foam or suds in carrying soil away from the work is not analogous to dispersion in the sense that it effects distribution of the soil throughout the detergent solution. However, the net result of "displacing the soil from the work" is the same. Much has been said of the role of foaming in this respect and, potentially, soil removal in the foam can be very significant. Under proper conditions, the solids- and oil-carrying capacity of bubble films can be very high, as attested by the well-known flotation processes. Not only is there a definite tendency for such materials to concentrate in the film and thus be lifted as the bubble rises through the solution, but their presence may increase the strength of the film and, thus, the stability of the resulting suds. This action has not been studied nearly so extensively in laundering as in flotation. Further, the efficacy of the action is probably considerably canceled by the design and method of operation of conventional washwheels. Certainly any operation wherein the suds with their load of entrapped soil are not permitted to remain entirely out of contact with the wash load cannot be expected to utilize fully the potentialities of foam as a soil remover.

13

Stabilization of Dispersed Soil

It is important in a discussion of the laundry washing process to differentiate between two very important opposing actions, namely, separation of soil from the fabric and redeposition of soil on the fabric. As will be evident from the discussions that follow, prevention of redeposition is equally as significant and probably as difficult to attain as complete separation of the original soil from the work.

The significance of redeposition of soil in the laundry sudsing operation may well be illustrated by experimental results obtained by Rhodes and Brainard,[1] as follows:

Pieces of white cotton material treated with a standard soil were washed five hours with untreated pieces of the same material. At the end of the wash period it was noted that all pieces showed the same brightness in a reflectometer, whether originally soiled or not. This brightness was intermediate between that of the original soiled and unsoiled pieces. In other words, soil was removed from the soiled pieces and redeposited on both the soiled and unsoiled pieces until an equilibrium was established between removal and redeposition. Repetition of the test, using the same soiled pieces from the first washing, showed pronounced increase in difficulty of removing soil, indicating that the redeposited soil was more tenaciously held to the fabric than was the original soil.

Although it is true that the washing period in the above-mentioned tests was much exaggerated in order that equilibrium might be

[1] Rhodes, F. H., and Brainard, S. W., *Ind. Eng. Chem.*, **21,** 60–8 (1929).

reached between the two opposing actions of removal and redeposition, the tests do indicate conclusively the existence of the two actions simultaneously.

Vaughn and coworkers[2, 3] differentiate between detergency and whiteness retention, assigning to the first the process of actual soil removal and to the second the prevention of redeposition of soil. They recognize the redistribution of soil between the detergent solution and the unsoiled and soiled pieces and conclude that, for soil to be simultaneously removed from and redeposited on a cloth, two opposing effects or a combination of effects must exist. Thus, in laundering, soil will continue to be removed from the cloth until the conditions tending to remove the soil are balanced by opposing conditions tending to redeposit soil on the same cloth.

Before entering into a discussion of prevention of redeposition by the detergent itself, brief mention will be made of the mechanical means by which this end is partially accomplished, namely, the use of multiple suds. Multiple suds are used in the washroom primarily on the premise that several short suds can accomplish equal or better soil removal with much less opportunity for soil redeposition. Thus, in the first suds, soil is removed from the work up to a certain point and in a short time; then the wash solution is "dumped" before the concentration of soil in the solution or time of exposure to the work is sufficient for an appreciable amount of redeposition to take place. A complete washing, consisting of a series of these short suds, thus reduces the opportunity for the occurrence of redeposition.

One-suds washing of soiled and unsoiled pieces simultaneously can be illustrated as in Figure 13–1, which is diagrammatic only. As the amount of soil removed from the soiled piece increases with time and, consequently, the concentration of soil in the wash solution also increases, redeposition of soil comes more and more into play. This redeposition is manifested not only by the decrease in whiteness of the unsoiled piece but also partly by the decrease in rate of increase of whiteness of the soiled piece. Thus, if all soil on the soiled piece were equally readily removed and if redeposition of soil onto the soiled piece were entirely prevented, the plot for whiteness *vs.* time of suds for the soiled piece would be a straight line as represented

---

[2] Vaughn, T. H., Vittone, A., Jr., and Bacon, L. R., *Ind. Eng. Chem.*, **33**, 1011–19 (1941).

[3] Vaughn, T. H., and Vittone, A., Jr., *Ind. Eng. Chem.*, **35**, 1094–8 (1943).

by the dashed line in Figure 13–1. When eventually the rate of removal of soil becomes equal to the rate of redeposition, the curves for the two test pieces converge and continue horizontally. In a one-suds washing and with all factors constant except the composition of the detergent solution, the vertical displacement of the two curves of Figure 13–1 will depend on the efficacy of the solution as both a soil remover and a preventer of redeposition. The more efficient the detergent in these respects, the higher will be the position of both curves at any given time.

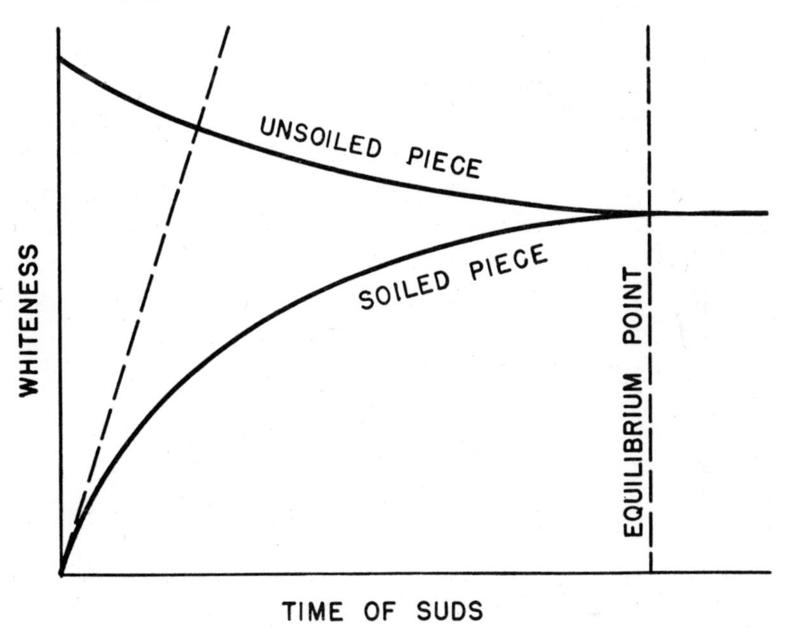

Figure 13–1. Diagrammatic representation of a one-suds washing.

The case for multiple-suds washing is well presented by the data of Rhodes and Brainard[1] shown in Figure 13–2. The curve for the 5-hour wash represents the data of their previously discussed 5-hour one-suds tests. The other curves are for five-suds washes, all with 0.25 per cent commercial soap solution at 40°C. The greatly increased efficacy of washing, both with the increase in number of suds and with the decrease in length of time of each suds (down to about 7 minutes per suds) is obvious from the curves. It is well to bear in mind, however, that these tests represent only one detergent solution and one method of washing, other than the number and time of suds.

They do not preclude the possibility of the development of a different washing technique that may accomplish the same end result with fewer suds. Thus, it is true that good commercial work, using multiple suds, accomplishes somewhat greater whiteness retention than good home laundry work, using only one suds. However, the magnitude of that difference is hardly sufficient to obviate a question as to whether or not present commercial washwheels and multiple-suds operation show the ultimate that can be expected in washing efficiency.

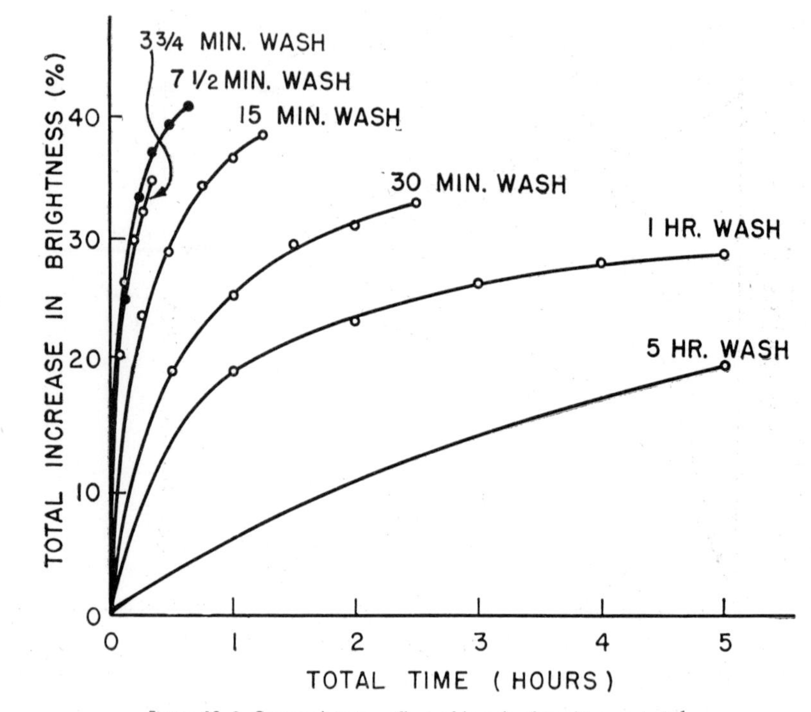

Figure 13–2. Curves showing effect of length of washing periods.[1]

Regardless of the mechanical technique that may be used to circumvent soil redeposition, of considerably greater and more direct importance is the role played by the detergent solution itself in preventing redeposition or, in other words, in so stabilizing dispersed soil that it does not readily separate out on the work. First, we must have the soil dispersed to such an extent physically that it is not readily entrapped in the fabric; next, the dispersed soil must be so stabilized that it does not collect into aggregates or droplets sufficiently large to become readily entrapped; and finally, there are other factors to be considered, such as establishment of a "repulsion" between fabric and soil.

The theoretical aspects of the stabilization of suspensions and emulsions by detergent solutions have been considered in Chapter 5. Thus, a dispersed body may be made to repel another dispersed body by imparting like electrical charges onto each. One dispersed body may be shielded from the influence of others by incasing it in a protective colloid which, in effect, surrounds it by an atmosphere of water of solvation. The size of the dispersed body, whether it be a solid particle or a droplet, is determined during the previous action of dispersion; the stabilizing agents serve only to maintain this previously determined degree of dispersion without improving on it. The efficacy of stabilizing agents in preventing redeposition in this respect is therefore seemingly limited, first, by the degree of dispersion and, second, by their ability to maintain this degree of dispersion.

The action of maintaining dispersion can hardly account, by itself, for all the power of a detergent solution to prevent redeposition of soil. Many of the interstices in fabrics through which the detergent solution "flushes" during washing must be sufficiently fine to catch or filter off even quite finely dispersed particles or droplets. An additional and perhaps very significant factor must come into play, namely, electrostatic repulsion between fabric and suspended soil. All the previous discussions of the manner by which an electrical charge is imparted to dispersed particles and droplets may be applied without reservation to the "charging" of the fabric. Further, the sign of the charge imparted to the fabric in an alkaline detergent solution is *negative*, as is the sign on the particles and droplets. It is entirely conceivable that electrostatic repulsion between fabric and soil aids greatly in prevention of redeposition, and even in the original separation of soil. This suggests an interesting and potentially valuable field of study.

14

Comprehensive Summarization of Detergent Action

Any over-all theory of detergent action in laundering, to be truly comprehensive, must give due consideration to all the significant individual actions involved in the detergent process and must apply to all usual conditions of soiling and washing. It must take cognizance of the functions of each component involved in a deterging system and must provide for the removal of all the soils that might be encountered in normal laundry work. None of the theories of detergent action advanced in the past meets such rigid requirements. A frequent fallacy in past theories has been a failure to duly recognize all significant factors. Such a commonly made statement as "Detergency is accomplished due to the ability of soap to lower surface tension and to serve as an emulsifying agent for oils," or words to that effect, is perfectly accurate as a *partial* statement of the *functions* of soap in detergency but it falls far short of being a definition of detergency.

Components of the Deterging System

We have discussed in considerable detail the specific functions of soap or synthetic detergents and of builders in detergency. We also have mentioned briefly the parts played by water itself and by mechanical agitation. It is appropriate that, in a comprehensive *summarization* of detergent action in laundering, the functions of all the components in the deterging system be reviewed. These components are:

a. Water;

b. Soap or synthetic detergent;

c. Builder;
d. Heat;
e. The load;
f. The soil;
g. Washwheel.

Functions of Water

Water is perhaps the most important single component in the system. Through its solvent action, water alone in many cases is an effective and complete detergent for a large percentage of the soil. Accompanied by suitable mechanical action, water may even serve as a complete detergent for all soil. It is our oldest detergent and the only self-sufficient one. Soaps, synthetic detergents, builders, and even "dry" cleaners, have been developed only for the purpose of improving the actions that water and mechanical work are capable of doing alone.

In conventional laundry practice, water serves not only as the solvent for water-soluble soil; it is also the wetting agent that penetrates the soil-fiber interface to replace the soil from the fabric, being assisted in this action by dissolved surface-active detergents such as soap. Water is the vehicle for carrying the detergent to the point where it is needed, and it is the vehicle for carrying away the separated soil. Water alone may effect appreciable dispersion of soil by peptization of aggregates of hydrophilic solid soil, such as albuminous or earthy material.

Lastly, water serves the very useful function of evenly distributing the mechanical action imparted by the washwheel. It applies directly to the soil and to the fabric, by flushing, considerable of the mechanical force essential for complete soil removal. By impact, water applies much of the force required to reduce particle and globule size for stable dispersion.

Functions of Soap or Synthetic Detergent

The most direct way of describing what is probably the *principal* function of soaps or synthetic detergents, *per se*, in detergency is to say that they serve to render hydrophobic surfaces hydrophilic. Because of the detergent becoming positively adsorbed not only in the water but also on the surface of both soil and fabric, there is formed in effect a "detergent-water" interface instead of a fabric-water or soil-water interface. As has been pointed out previously in Chapter 10, the simplest way of eliminating a soil-fabric interface is to substitute another interface of lower free interfacial energy. Such substitution is effected when detergent is adsorbed on fabric and on soil to produce

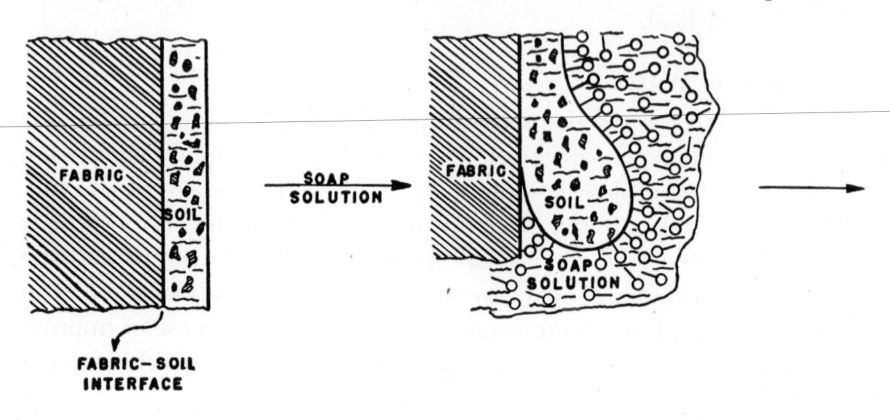

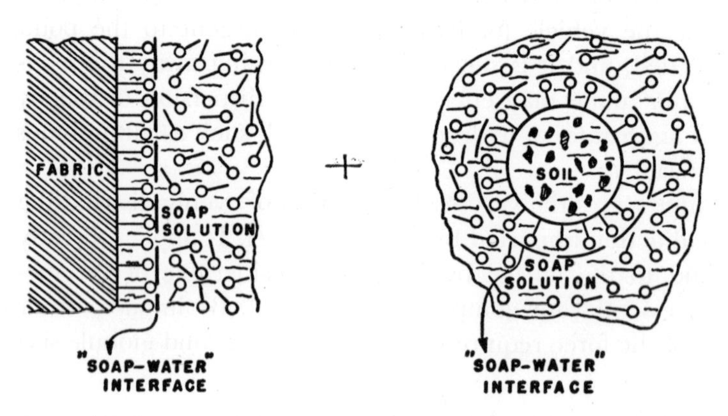

Figure 14–1. Illustration of the role of detergents in substituting hydrophilic interfaces for fabric-soil interfaces.

two new "detergent-water" interfaces in lieu of the original soil-fabric interface. This substitution of interfaces is illustrated schematically in Figure 14–1.

Rendering the soil more hydrophilic through the use of soaps or synthetic detergents, as described above, aids the water and applied mechanical action not only in removing soil from fabric but also in dispersing the soil and in maintaining it in suspension. The reduced free surface energy of the soil reduces the mechanical work necessary to attain a desired degree of dispersion of aggregates. The increased hydrophilicity of the dispersed soil particles or droplets permits greater stability in aqueous suspension and lesser tendency to re-aggregate and redeposit on the cloth.

Another function of soaps, particularly in the absence of alkaline builders, is to serve as a source of alkalinity for the neutralization of acidic material in soil. This function is fulfilled little if at all by synthetic detergents because of the practical absence of hydrolysis.

These and other functions of soaps and synthetic detergents may be summarized as follows:

a. Surface and interfacial tension depressants:
   (1) Improve the wetting power of water;
   (2) Permit more complete and rapid penetration of water into and around the soil and fabric;
   (3) Reduce the mechanical energy required to effect separation of insoluble soil from fabric.

b. Solvents:
   (1) Fatty portion of the detergent molecule is a solvent for some organic soils;
   (2) Detergent micelles may be solubilizers for both liquid and solid soil;
   (3) Act as introfiers (mutual solvents) for both water and oil.

c. Dispersing agents:
   (1) Enhance the ability of water alone to disperse solid aggregates by peptization;
   (2) Reduce the mechanical energy required to effect dispersion of both solid and liquid soil.

d. Stabilizing agents:
   (1) Increase the stability of dispersed soil in water (reduce the tendency to flocculate or re-aggregate) by increasing the hydrophilicity of the dispersed material and, thus, its solvation;
   (2) Furnish controlled electrolytic stabilization for dispersed soil.

e. Source of alkalinity (soaps only):

(1) For neutralizing acid soil and, possibly, saponifying fatty material.

f. Supply foam (except certain non-foaming synthetic detergents):

(1) To assist in carrying soil away from the fabric.

The above list of functions of soaps and synthetic detergents may be considered complete in so far as it covers functions which have been recognized and considered in past studies of detergency. Past attempts to assign to each function its relative "worth" in the over-all process of detergency have been unsuccessful. This list of functions has one serious shortcoming which is readily apparent when one considers that it does not reveal the reasons why some highly surface-active agents are not good detergents, or why soaps are better than synthetic detergents in some cases while the reverse is true in other cases. This indicates that either the list of functions is incomplete or that the relative extent to which a particular surface-active agent fulfills the various functions determines its efficiency as an over-all detergent.

Functions of Builders

Because of their ability when properly used to enable "less detergent to do more deterging," and in view of their generally lower cost, builders are truly detergent extenders. They frequently permit the attainment of better end results, in terms of both better quality of finished work and lower cost, than is possible when unbuilt solutions are used. In serving thus, they fulfill specific functions which may be summarized as follows:

a. Enhance the interfacial activity of soaps and synthetic detergents at a given detergent concentration;

b. Neutralize acidity in the soil (alkaline builders only) to conserve soap and to render acid soil more "soluble";

c. Partially act on saponifiable fatty soil (alkaline builders only) to render it more soluble;

d. Inhibit hydrolysis of soaps to fatty acids or acid soaps (if sufficently alkaline);

e. Enhance spontaneous emulsification by "on-the-spot" formation of soaps from fatty acid soil (alkaline builders only);

f. Electrolytically stabilize emulsions and suspensions of soil in some cases;

g. Serve as protective colloids for stabilization of soil suspensions in some cases;

h. Enhance foam formation and foam stability;

i. Enhance electrical repulsion between fabric and soil.

As was the case with soaps and synthetic detergents, this list of functions of builders is insufficient in the sense that it does not furnish the basis for explaining why one builder may serve better than another for a particular application.

Functions of Heat

Heat, although less tangible, is as much a component of a deterging system as is, say, water or soap. It is usually added in measurable quantities to aid in the detergent process. Its use entails a definite cost per unit quantity, as does the use of soap involve a definite cost per pound, and must be justifiable on the basis of ability to fulfill certain functions as follows:

a. Of perhaps greatest importance, heat and the consequent higher temperatures of the deterging system permit the use of soaps and synthetic detergents of higher molecular weight than would otherwise be possible; this in turn permits the effective utilization of a broader source of cheaper materials;

b. In connection with the foregoing, heat may increase the solubility of the detergents of higher molecular weight to the point where the concentration of maximum surface activity is attainable (on the other hand, it must be pointed out that an increase in temperature *decreases* the surface activity of solutions of detergents which are soluble to the optimum concentration in the cold);

c. Heat enhances the spontaneous separation of soil from fabric by:

(1) Decreasing the strength of the adsorptive bonds between soil and fabric;
(2) Decreasing the viscosity of liquid soil and, thus, the resistance of the soil to shear during dispersion;
(3) Increasing pedesis of solid soil particles;
(4) Increasing the solubility of soluble soils;
(5) Increasing the rate of reaction and, thus, the extent of reaction in a given time between alkali and acid soil or saponifiable fatty soil.

In contrast to the above-listed beneficial functions of heat there are many ways in which heat may be detrimental to the efficiency of a deterging system. This combination of beneficial and deleterious actions probably is more pronounced in the case of heat than in the case of any of the previously discussed components, water, detergent, or builder. Thus, we have among the deleterious actions of heat: the lowering of the surface activity of solutions of readily soluble detergents; increase in the extent of hydrolysis of hydrolyzable detergents; decrease in stability of emulsified liquid soil; decrease in solvation of dispersed soil particles; increase in settling tendency of suspended solid particles. In addition, there probably exist definite temperature coefficients for the zeta potentials of various dispersed materials, with consequent optimum temperatures for maximum potential and, thus, maximum electrolytic stabilization. It is obvious from the foregoing that the decision as to whether or not applied heat is to be used in a detergent process and as to the temperature at which the deterging system shall be maintained must depend on the "balance" between beneficial and deleterious effects that may be expected under the conditions of the particular application.

Functions of the Load

It does not require too great a stretch of the imagination to realize that the load itself, that is, the fabrics that are being washed, may contribute to their own deterging. Specifically, the load applies directly to the soil as a rubbing and "bumping" action considerable of the mechanical force imparted by the washwheel and essential for complete soil removal and for adequate soil dispersion. Proof that the load fulfills a definite function in this respect is obvious from the well-known fact that successful washwheel operation is very dependent on the manner in which the load is placed in the wheel and, thus, on the manner in which the load may rub and "bump."

Functions of the Soil

The ability of certain components of soil to aid in the deterging of other components amounts in effect to the soil "lifting itself by its own boot straps." The functions of fatty and fatty acid soils, after conversion to soaps by alkali, are the same as those of added soap, with the further advantage that they are immediately disposed where they may be particularly effective. In addition, certain soils adsorbed in suds films may strengthen the films and thus enhance their soil-carrying capacity.

Functions of the Washwheel

In addition to serving as the container for the detergent process the washwheel may be considered as the embodiment of the mechanical action essential to complete detergence. The chief function of the washwheel is to impart the mechanical action. In turn, the functions of mechanical action may be enumerated as follows:

a. Furnish much of the energy desirable for aiding in the reduction of the adsorption bonds between soil and fabric and, thus, for aiding soil separation;

b. Furnish most of the energy essential to dispersion of soil into smaller particles or droplets, the magnitude of this energy being measured by the increase in free surface energy attending the dispersion of the particles and droplets;

c. Furnish most of the energy required to maintain all except extremely small particles in suspension, that is, the energy required to overcome gravitational effect and to overcome the tendency of fabric to filter soil out of suspension. (This energy must be continuously supplied to the system for such a time as the suspension is maintained.)

By far the greatest amount of energy which must be supplied to a deterging system is applied as mechanical energy, the only other sources being applied heat and energy evolved from such chemical reactions as take place in the system. It is not within the scope of the present discussion to review washwheel design and methods of operating washwheels for the purpose of determining their efficacy as sources of mechanical energy.

Surface Activity vs. Detergent Power

The surface activity of soaps and synthetic detergents often has been considered as the ultimate source of their detergent power. In turn, measurements of the surface activity of such materials have been taken frequently as true measures of detergent power. Such practice demands particular consideration of the relative significance of surface activity in the over-all detergent process of soil removal, soil dispersion, and prevention of redeposition.

Preston[1] has ably discussed several correlating principles of detergent action and has compiled from several sources data such as that shown in Figure 14–2. These data show that, within a narrow concentration range, some striking change occurs in solutions of a detergent to produce decided changes in many physical characteristics. This range is readily associated with the critical concentration for formation of micelles. Preston shows that micelle formation in

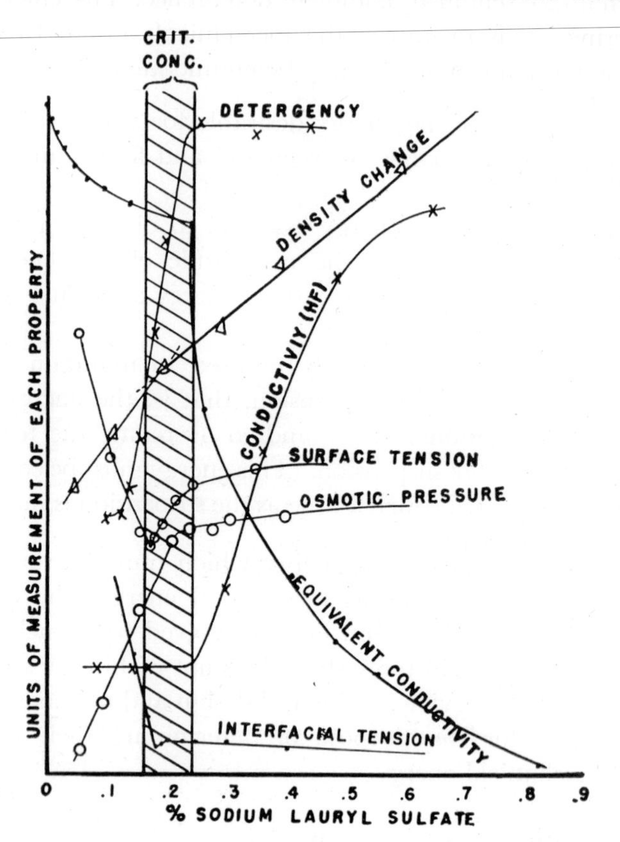

Figure 14–2. Physical property curves for sodium dodecyl sulfate.[1]

dilute solutions of surface-active agents is simply another manifestation of the surface activity of the solute, just as is the lowering of surface tension or interfacial tension. Thus, the high free energy at the interface between solute and water may be reduced by expelling the solute to the surface of the water or by aggregation of the solute into

[1] Preston, W. C., *J. Phys. & Colloid Chem.*, **52**, 84–97 (1948).

micelles of lower total surface area. In view of the fact that micelle formation in dilute solution comprises an aggregation of ions, such aggregation can occur only when the forces promoting the lowering of free surface energy exceed the repulsion between ions of like charge.

As shown in Figure 14–2, the maximum detergency *of a detergent* is attained in the *same concentration range which gives maximum surface activity*. The "breaks" shown in the figure for the other properties also occurring in this critical concentration range, namely, in density change, conductivity, and osmotic pressure, are simply further manifestations of micelle formation and have little, if any, bearing on detergency.

The work of Preston affords several very interesting and valuable insights into detergent action. Based on extensive studies of washing action *vs.* both detergent composition and concentration, wherein the proportion of load to detergent solution was purposely *low* so as not to affect greatly the concentration of free detergent in solution, Preston has been able to present the following summarizations:

a. For any one of the sodium soaps, laurate, myristate, palmitate, and stearate, the concentration at which the increase in washing power with increase in concentration begins to fall off corresponds approximately to the concentration at which deflections are found in the curves for conductivity, pH, density, and surface activity. If the latter deflections are to be attributed to colloid formation, it is reasonable to attribute the deflections in washing power to the same cause.

b. From the curves for washing power *vs.* concentration for sodium and potassium soaps and for sodium alkyl sulfates and sulfonates, critical washing concentration can be plotted against the number of carbon atoms as shown in Figure 14–3. It is seen that the points for all four groups of detergents fall close to a smooth curve, as if differences in the polar heads ($—COO^-$, $—SO_4^-$, and $—SO_3^-$) have slight effect, compared with chain length.

c. ". . . the molar critical washing concentration is essentially the same for the sodium as for the potassium soaps, and . . . the detergency curves for sodium, ammonium, magnesium, copper, and triethanolamine alkyl sulfates all show breaks at about the same molar concentrations. . . . This would be surprising if the active washing constituent were either the molecule or a colloidal micelle containing much cation, but it is not surprising on the basis of the

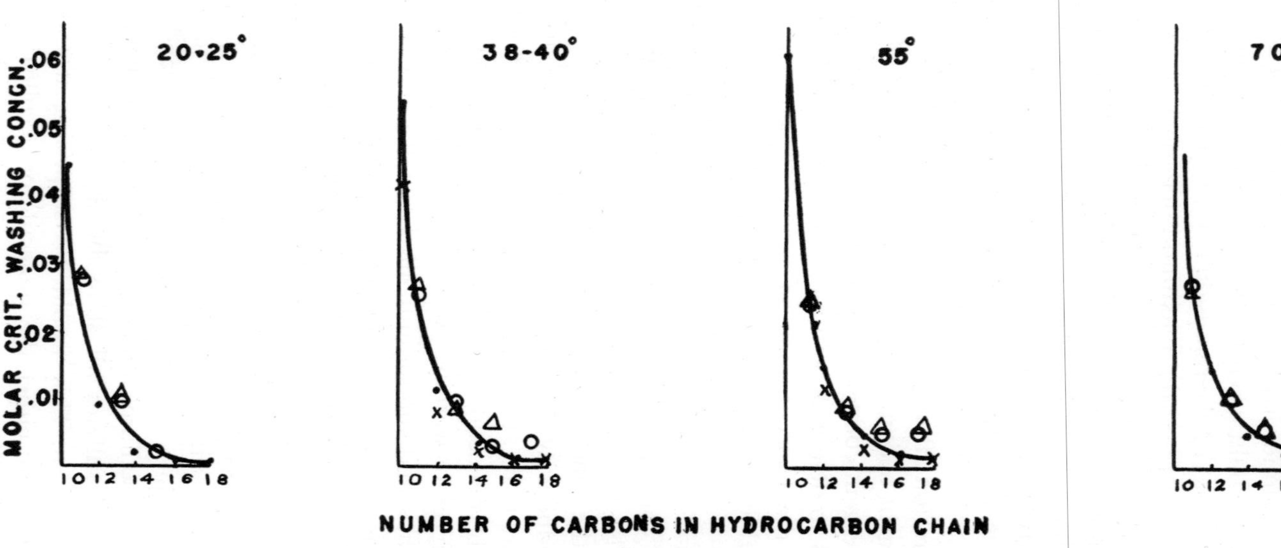

Figure 14–3. Correlation between critical washing concentration and length of hydrocarbon chain.[1]

×—sodium alkyl sulfates.
●—sodium alkyl sulfonates.
○—sodium soaps.
△—potassium soaps.

concept here proposed, the concept of the importance of the long-chain ion which would of course be the same whether derived from a sodium salt or from any other salt."

d. "The hypothesis that the long-chain ion is the active constituent of detergent solutions is applicable to both anionic and cationic detergents. In the case of the non-ionized type, the molecule itself would be the active constituent."

e. "Detergent action, colloid formation, and surface activity are different manifestations of the same characteristic of the detergent. The long-chain ion (for ionic detergents) is the active constituent in each manifestation. Colloid formation begins, and washing power and surface activity reach their maximum, at the concentration at which further additions of detergent either (a) do not dissolve (at low temperature) or (b) dissolve to form colloid (at higher temperature), and thus in neither case is further increase in the number of long-chain ions in the solution possible."

At this point it should be mentioned again that Preston used in his studies of washing power small loads of soiled cloth in relation to volume of detergent solution. It would appear from this that his postulation as to the surface activity of fatty ions being the "answer" to detergent power might necessarily be limited to soil removal and soil dispersion. From the fact that much if not all of the tendency for soil to redeposit during his tests might have been due to the high relative volume of solution, he has little experimental substantiation for inclusion of soil stabilization in his postulation. The case for surface activity being the answer to the question—"What is the source of the power of detergents to aid in soil removal and dispersion?"—is quite complete from his and other data. On the other hand, the case for it being the source of power to prevent redeposition is not so clear. Preston[2] believes that "the properties of a solution which enable it to remove soil particles from a cloth will also enable it to prevent those particles from again attaching to the cloth." Certainly such properties are at least partly the answer to prevention of redeposition; however, the sufficiency of this parallelism as a complete explanation of stabilizing power lacks confirmation.

Granted from the foregoing discussion that surface activity, *per se*, is the prime motivating force possessed by detergent solutions

[2] Preston, W. C., Private communication.

for at least the accomplishment of soil separation and soil dispersion, there remains to be determined the answer to the ever-present question as to why surface activity does not *always* determine detergent power. Certainly surface activity as manifested by lowering of surface and interfacial tensions is not the final answer. Specifically, we have first the case of agents whose surface activity (as measured by surface- and interfacial-tension lowering) is quite comparable to that of detergents but whose over-all detergent power is poor. Finally, we have the case of many agents whose apparent surface activity is inferior to that of true detergents but which are equal to the detergents in certain single actions such as hastening wetting or aiding emulsification. Presumably a material not only must be surface-active to be a detergent but it must also be able to put this surface activity to use in a peculiar combination of ways. This enigma is probably the most important and the most difficult of the several that remain to be answered in the realm of detergency.

15

Present Knowledge of Detergency vs. Present Washroom Practice

Knowledge is sought and acquired for either of two purposes: (a) to satiate man's liberal curiosity; (b) to serve some utilitarian function, either directly or as an aid to the attainment of further more-practical knowledge. Certainly in the present discussions we are concerned primarily with the utilitarian aspect of our knowledge of detergency. Therefore, we should examine the manner in which that knowledge is being put to practical use in laundering.

The detergent process in laundering has been defined previously as that part of conventional laundering carried out in the washwheel up to the point of rinsing and bleaching. As such it should include pre-soaking, whether accomplished in the washwheel or elsewhere.

Cold-water pre-soaking is not practiced generally in commercial laundering, although quite commonly in home laundering. In lieu thereof, it is the practice to run the first or "break" suds operation in the relatively low temperature range of 100°–125°F, with sufficient built detergent to maintain a high suds level throughout a washing time which may be as much as ten minutes.[1,2] Such temperatures are sufficiently low that thermal coagulation or "setting" of albuminous soil in the fabric is prevented. Denaturation of albuminous soil prob-

---

[1] American Institute of Laundering, "Washing Formulas," Technical Bulletin No. 34, Joliet, Illinois.

[2] Procter & Gamble, "Better Laundering," Ivorydale, Ohio (1942).

ably occurs in the presence of the alkalinity of the wash solution; however, the denatured protein thus resulting should remain soluble so long as the alkalinity is maintained.[3] Quantitative determinations of soil removal by Bray[1] indicate approximately 22 per cent soil removal in the first operation with water alone, 31 per cent with water and alkali, 40 per cent with neutral soap, and over 50 per cent with a heavy suds obtained with proper proportions of soap and alkaline builder. Such obvious advantage in the first operation from the use of built detergent plus the advantage of a temperature as high as possible (yet below the coagulation temperature), to simplify heating to a higher temperature in the second suds, indicate that the theoretical potentialities of cold-water pre-soaking have been improved upon in commercial practice.

The amount of soap to be added to the wash bath is measured in terms of the depth of running suds which it will maintain in the wheel. This is not simply a carry-over from the period when the suds were considered to be the all-important soil remover. So long as adequate suds are maintained, the operator knows that there is no deficiency in the amount of dissolved soap. Disappearance of the suds during a washing operation is a danger signal indicating consumption of available soap to the extent that the wash solution no longer can work effectively. The soap demand of a particular load is determined by such factors as amount of the load and extent of soiling and may be quite variable between loads. The operator therefore requires a "gauge" which readily and quickly indicates the sufficiency or deficiency of soap. The depth of running suds serves such purpose and no better indicator has been developed. However, such a method has certain inherent deficiencies. In the first place, there may be quite an appreciable variation in actual available soap content of the wash solution with no appreciable variation in depth of running suds. Secondly, it was only by trial and error that a certain depth of suds was decided upon for a particular sudsing operation; we have no direct confirmation that a certain depth of suds definitely denotes optimal detergent power. Finally, depth of suds may give only "loose" control over the addition of excessive amounts of detergent, even though such excessive amounts may be detrimental rather than bene-

[3] Harrow, B., "Textbook of Biochemistry," 3rd ed. (revised), p. 51, Philadelphia, W. B. Saunders Company, 1944.

ficial to both the efficiency and the economy of the process. Attempts have been made in the past to overcome these deficiencies by development of a better "gauge" (without success) and by closer manual control of suds depth (with considerable improvement). There would seem to remain considerable room for development of controls that more definitely will assure maximum detergent efficiency with minimu m wastage of supplies.

The practice of multiple-suds operations already has been introduced (page 220). If it were not for the definite tendency of soil to redeposit, there would be no theoretical justification for multiple-suds washing in lieu of single-suds washing. This practice is only one of two general ways in which redeposition is being inhibited, the other being stabilization of the soil dispersed in the wash solution. There is no question but what the necessity for multiple suds results in a considerably more costly operation than would be the case for single suds. The cost of detergent, water, heat, power, and capital investment that would be involved in single-suds operation is increased in multiple-suds operation by a factor approaching the number of suds necessary to accomplish complete washing. Certainly the ultimate goal for development both of detergents and of washwheels in this respect is detergents and washwheels of such *combined* efficiency that a single-suds operation will produce the desired results. Detergents are being continually developed toward the goal of more effective stabilization of soil when used in presently available washwheels. On the other hand, washwheels are being developed toward more effective use of presently available detergents. A coordination between these two lines of endeavor is lacking. Neither detergents nor the equipment in which they are used can be considered completely developed until they jointly permit single-suds operation. For the final attainment of this ultimate goal it is necessary that our knowledge of the basic factors involved in physico-chemical stabilization of dispersed systems and in the mechanical design of washing equipment be considerably advanced.

Undoubtedly the most difficult problem faced by the laundryman in his washroom operations is that arising from the varied natures of the materials to be washed and the varied susceptibilities of different fabrics to deleterious action by detergent solutions. Complicated systems of classification are required, wherein each classification receives special handling in the washroom.[4] The factors governing classification are, briefly:

a. Chemical resistance of the fabrics to detergent solutions;
b. Color-fastness;
c. Mechanical strength of the fabrics, particularly when wet;
d. Nature and extent of soiling;
e. Bleach resistance of the fabrics and their dyes.

These factors frequently determine the maximum temperatures, maximum alkalinities, and the nature and extent of mechanical action permissible in washing, even though these maxima may not always conform to the temperatures, alkalinities, and mechanical action that give maximum detergent efficiency.

When the nature of the work dictates the use of low temperatures and low alkalinities, as is the case with woolens, the laundryman selects those detergents which are most effective under such condtions. When the nature of the work permits higher temperatures and higher alkalinities, as is the case with white cottons, a different choice of detergents is made. In this way, the basic knowledge of the influences of composition, concentration, and temperature on the effectiveness of detergent solutions is put to practical use. However, such manipulations in supplies and operation are not the *solution* to the difficulties imposed by complex classification. Rather, they are a means of circumventing the difficulties and they add much to the cost and complexity of laundering. The ultimate solution to the difficulty would be the development of such washing techniques that complex classification would be unnecessary. Paramount among the requirements of such techniques are low temperature and low alkalinity. In this respect unhydrolyzed synthetic detergents show great promise. They offer the only presently apparent avenue of approach to the ideal universal wash formula for all work.

The foregoing brief discussion indicates that generally good use is being made of our presently available basic knowledge of detergency. The discussion also indicates certain directions in which future basic studies might be pointed to advantage.

In closing, brief reference will be made to special laundering techniques developed by the British Launderers' Research Associa-

[4] American Institute of Laundering, "Classification," Service Bulletin No. 2, Joliet, Illinois.

tion but not used in this country.[5] The first step in this process is carried out in soap solution in the absence of alkaline builder. Omission of builder is predicated on the known effect of alkalies in raising the surface tension of soap solution and the deduction therefrom that more rapid wetting will be realized in unbuilt solutions. The second step in the process is a remarkable departure from conventional American practice in that the pH of the wash solution is reduced below the point of decomposition of soap (to about pH 8.5) for a short time. This reduction of pH results in decomposition of the soap to fatty acid and, perhaps, to acid soap, materials which are particularly active toward oily soil and thus intermingle with the soil. The third step in the process involves raising the pH of the wash solution to about 11.0 by the addition of a suitable alkaline builder. This results in conversion of the fatty acid and acid soap *in situ* back to soap, the latter now being in a position with respect to the soil where it is particularly effective in accomplishing soil removal. Harwood reports savings in soap of as much as 40 per cent and savings in time of as much as 30 per cent by this method over older methods. Although these techniques are not known to have been studied in this country, their apparently successful use in Great Britain over a period of several years would indicate that they serve as an excellent example of what can be accomplished by concerted research effort, properly directed.

[5] Harwood, F. C., *J. Textile Inst.*, **39,** 513–25 (1948).

Appendix

The Theory of Cassie and Palmer[1] on the Effect of Electrolytes on Surface Activity

The process of positive adsorption of surface-active solute molecules or ions in a surface or at an interface involves a reduction in the total potential energy of the system. The efforts of the detergent to enter the surface are opposed by electrostatic repulsion from detergent ions already in the surface.

The Maxwell-Boltzmann distribution law relative to the proportion of the total number of molecules in a system whose energies are greater than some value E (expressed in calories per mole) may be expressed approximately as

$$\frac{n'}{n_t} = e^{-E/RT} \tag{1}$$

where n' is the number of molecules having the greater energies, n_t is the total number of molecules in the system, R is the gas constant, and T is the absolute temperature. Cassie and Palmer have applied the expression of equation (1) to the conditions in a solution of strongly electrolytic anionic detergent as

$$\frac{c_s}{c} = e^{\omega/kT} \tag{2}$$

where c is the bulk concentration of solute in fatty anions per cc, ω is the decrease in potential energy of the system when a fatty ion passes from the bulk into the interface, c_s is a measure of the surface con-

[1] Cassie, A. B. D., and Palmer, R. C., *Trans. Faraday Soc.*, **37**, 156–68 (1941).

centration in anions per cc, and k is the gas constant per ion (1.3805×10^{-16} erg/degree). It will be seen that, in a sense, equation (2) is a reciprocal of equation (1); rather than expressing a proportion of ions of potential energies *greater* than a set value, equation (2) expresses a proportion of ions (those adsorbed in the interface) of potential energies less than the average potential energy of the ions remaining in the bulk of the solution.

The first assumption by Cassie and Palmer is that c_s, a value that cannot be easily interpreted because of its dimensions (number per cc), is a function of only the thermodynamic (Gibbs) surface excess, Γ, expressed as number of anions per sq. cm. Then

$$c_s = c_s(\Gamma) \tag{3}$$

A second assumption, to simplify the derivation, is that a constant surface pressure (constant surface or interfacial tension) implies a constant excess of fatty anion in the interface, or that for one fatty anion at one interface the surface or interfacial tension is a function of the surface excess only and is dependent on the ionic distribution below the surface film only by the effect of such ions on the surface excess of fatty anions.[2] A third assumption made by Cassie and Palmer is that positively adsorbed ions will produce effectively a plane charge at the interface.[3] When an ion of the same sign as that of the ions already adsorbed in the surface passes from the bulk of the solution into the surface, the potential energy of the system will

[2] This assumption appears to be quite plausible. However, for it to be strictly accurate, any tendency for gegenions to pull the fatty end of the surface anions more deeply into the solution must have no appreciable influence on the ability of that surface anion to lower the surface or interfacial tension. Experimental verification of the assumption would be a significant contribution to the answer to the question: As between an unbuilt and a built solution of the same surface-active agent, adjusted to the same surface tension (or interfacial tension) by variation in concentration of the agent, is the surface concentration of fatty anion the same for each solution; in other words, is the ability of any one detergent ion to lower the surface or interfacial tension, so long as it remains adsorbed in the surface, independent of the influence of gegenions?

[3] It should be pointed out in connection with this assumption that one of the conditions imposed by Cassie and Palmer for the "workability" of their mathematical treatment is that the over-all detergent concentration shall be sufficiently low that *concentrations* are reasonable approximations of *activities*. Minimum equilibrium interfacial tensions are attained only at concentrations where the interface probably is saturated with detergent. At the limited concentrations set by the hypothesis the interface may be quite unsaturated, with a considerable number of water molecules interspersed between the detergent ions at the surface. The accuracy of the assumption that the adsorbed anions produce a *plane* charge at the interface must depend on the required extent of saturation of the interface in order for the anions no longer to act only as a number of *point* charges.

be diminished, by the transfer, to a considerably less extent than would have been the case had there been no previously established plane charge. In other words, the first fatty anion to be adsorbed in a new interface effects a reduction in potential energy of the system which is greater in magnitude than the reduction effected by any subsequently adsorbed anion; the electrostatic repulsion between anions already adsorbed and anions just newly entering the interface reduces the magnitude of the decrease in potential energy effected by the newly adsorbed anions.

The change in electrical energy effected when an anion passes from the bulk of the solution into an interface is a function of the potentials existing in the plane of the film and at the point of origin in the bulk. If these potentials are designated as ϕ and ϕ', respectively, the change in electrical energy (a measure of the *work* to effect the transfer) is $z\epsilon(\phi - \phi')$, where z is the valence of the anion and ϵ is the charge on one electron (4.803×10^{-10} electrostatic unit). Cassie and Palmer hypothesize that the nonelectrical change in potential energy, ω, and the change in electrical energy are additive to give the total energy change of the process. Then, equation (2) becomes

$$\frac{c_s}{c} = e^{\omega - z\epsilon(\phi - \phi')/kT} \tag{4}$$

In any actual solution of electrolytic detergent, whether built or not, ϕ will always have a finite value; there will always be an atmosphere of gegenions to prevent ϕ being infinite. The potential, ϕ', at the point in the bulk of the solution, may be considered to have a value corresponding to the work involved against ϕ in bringing an anion from an infinite distance away up to the final distance of the point from the film. In accordance with the Debye-Hückel theory,[4] the activity of the ion in the bulk of the solution is a function of ϕ'. By replacing the concentration, c, by the activity, γc (where γ is the activity coefficient), ϕ' may be removed from equation (4) to give

$$\frac{c_s}{\gamma c} = e^{(\omega - z\epsilon\phi)/kT} \tag{5}$$

[4] See Glasstone, S., "Textbook of Physical Chemistry," 2nd ed., New York, D. Van Nostrand Company, Inc., 1946, for a comprehensive mathematical treatment of the Debye-Hückel theory.

A final hypothesis by Cassie and Palmer concerns the value of ω. They assume that $\Delta\omega$, which is the extent to which the decrease in potential energy upon adsorption is affected by one more or one less CH_2 group in the fatty anion, is approximately constant from 12 to 18 carbon atoms. They further assume that $n\Delta\omega$ is approximately equal to ω, where n is the number of CH_2 groups in the fatty chain. By applying available data to equations (2) and (3) they arrived at a value for $\Delta\omega$ of approximately $0.85kT$ for the primary long-chain anion, dodecyl sulfate, at a water-xylene interface. For films of adsorbed fatty anions of 12 to 18 carbon atoms, ω will be of the order of $12kT$, which value is subject to the accuracy of the assumption discussed by footnote (2), page 243. (At 300°K, this corresponds to approximately 5×10^{-13} erg per ion adsorbed.)

By referring again to equation (5), it may be seen that for the electrical effect, $\epsilon\phi$, to have much effect on the adsorption of an anion the term $\epsilon\phi$ must be of the order of magnitude of ω. Cassie and Palmer arbitrarily set a minimum significant value for $\epsilon\phi$ of $5kT$.

The density of electricity at a point is equal to the excess negative electricity per unit volume at that point. The relationship between electrostatic potential, ϕ, and charge density, ρ_e, at that point can be expressed according to the Poisson equation as

$$\frac{\delta^2\phi}{\delta x^2} - \frac{\delta^2\phi}{\delta y^2} - \frac{\delta^2\phi}{\delta z^2} = \frac{-4\pi\rho_e}{D} \tag{6}$$

where x, y, and z are the rectangular coordinates of the point in space and D is the dielectric constant of the water.[4] For the special case of a plane charged surface in contact with a solution of strong electrolyte, equation (6) can be converted to polar coordinates and integrated[5] to give

$$\frac{d^2\phi}{dx^2} = \frac{-4\pi}{D} \sum z_i n_i \epsilon e^{-z_i\epsilon\phi/kT} \tag{7}$$

where x is the direction normal to the surface and z_i is the valence of ions of number n_i per cc in the bulk of the solution. Cassie and Palmer then integrate equation (7), considering that when $x = \infty$, $\phi = O$ and $\frac{d\phi}{dx} = O$, to give for the boundary conditions

$$\frac{1}{2}\left(\frac{d\phi}{dx}\right)^2 = \frac{4\pi kT}{D} \sum n_i(e^{-z_i\epsilon\phi/kT} - 1) \tag{8}$$

[5] Müller, *Cold Spring Harbor Symposia on Quantitative Biology,* **1,** 3 (1933).

If the adsorbed film is considered as a plane surface of uniform charge, σ, per unit area,

$$\left(\frac{d\phi}{dx}\right)_{x=0} = \frac{-4\pi\sigma}{D} \tag{9}$$

Substitution into equation (8) gives

$$\sigma^2 = \frac{DkT}{2\pi} \sum n_i(e^{-z_i\epsilon\phi/kT} - 1). \tag{10}$$

(For further consideration of this equation, see p. 157, *et seq.*)

Author Index

Subject Index